I0402568

SIR ISAAC NEWTON AND ALBERT EINSTEIN

From Absolutism to Relativity

THE HISTORY HOUR

HISTORY

CONTENTS

SIR ISAAC NEWTON

ALBERT EINSTEIN

SIR ISAAC NEWTON

One of the Greatest Minds of All-Time

Genius is patience.

— SIR ISAAC NEWTON

INTRODUCTION

❦

Sir Isaac Newton was one of the finest minds to have ever lived. He was born on the old calendar on Christmas day in 1942, but in the new style of calendar, his birthday would be placed on January 4, 1643. He was born in Lincolnshire in England.

❦

He ended up being a prominent figure in many different aspects of life, but his main body of work was as a physicist and mathematician. He was a part of the scientific revolution in the 17th century, which would fundamentally change the way that people would see the world. In the field of optics, he advanced our understanding of light and how we saw it. In mechanics, he would create his famous three laws of motion, but it's in physics that he became most well-known for his understanding of gravity and in mathematics for his discovery of calculus, and his writing is perhaps the single most important scientific book of all-time, the '***Principia***,' which is still referenced today.

✣ II ✣
BORN INTO TRAGEDY

"We build too many walls and not enough bridges."

— SIR ISAAC NEWTON

❁

Isaac Newton's life was tragic before the young genius was even born into the hamlet of Woolsthorpe. By the time Newton was born, his father had already passed away three months before. As a beautiful moment of passing the torch from one to the next, Newton was born in the same year that Galileo died, and it could be said that Newton picked up where the Italian had left off.

❁

Newton was born as a small, premature baby who wasn't given a great chance to live. Born without a father, he would soon also be without a mother when he was three; his mother remarried and abandoned Newton to be raised by his grandfather. The tragic loss of his father and then the abandonment by his mother would clearly play a key role in some of the personality traits that he would develop.

※

For the next 9 years, he would be separated from his mother until his grandmother died. Newton resented his mother and hated his stepfather. He once wrote in his sins that he once threatened to burn them and their house. Newton would go on to be incredibly anxious and irrational, and these traits would stay with him throughout his life.

※

From the age of 12 to 17, Newton spent time living with a man who provided him lodgings while he studied at King's School in Grantham. This would be the point where Newton would first develop a real interest in science as he developed a fascination with chemistry. You'd think a man who would go onto become one of the greatest minds of all-time would have done well in school, but he was more interested in knowledge outside of the classroom and wasn't a great student.

※

At this time, he was removed from school, and his mother made an attempt to make him a farmer, but it was never going to work as Newton hated farming. The master at King's

School though must have seen Newton's talent, and he persuaded his mother to send him back to school. His mother agreed, and Newton's academic performance increased, and he completed the year with an impressive final report.

HIS LIFE IN CAMBRIDGE

꧁꧂

In 1661, Newton would find himself in Trinity College in Cambridge after excelling at his grammar school. He would still find himself in the world of mathematics, but his education would soon take a different path. Due to the delay in him getting to university, due to looking after his mother's land, Newton was slightly older than most of the undergraduates in the college.

꧁꧂

At the time Newton set foot in Cambridge, there was already a revolution going on in the world of science. The likes of Copernicus and Kepler had already proven that the sun is at the center of the solar system, and Galileo had laid the groundwork for understanding the laws of inertia. He was in an age where the world was discovering that we weren't the center of the universe. This though wasn't an easy process, and the likes of Cambridge were late in accepting this line of thinking as they still got most of their funding from the church.

❦

This line of thinking had come from Aristotle, and that's who Newton studied for much of his early years there. Whilst Aristotle was a great mind himself, his views on the universe were outdated as technology moved on from a man who had died in 322. While Newton took onboard a lot of his work, it was time to move on with some of these beliefs, but institutions like Cambridge were reluctant to do so.

❦

It was only when Newton furthered his reading that he began to understand other ways of thinking. He read works by the likes of Rene Descartes along with other philosophers who disagreed with Aristotle's view of the world. These were people who proposed that the world was made out of particles in motion and that nature was simply a result of their mechanical interaction.

❦

It was at this stage that Newton would search for the truth and start a notebook with a slogan that read

"Plato is my friend, Aristotle is my friend, but my best friend is the truth,"

which is a very scientific approach where everything is questioned until it's proved. It set the basis for a life where

Newton would have his ideas and then find a way to prove them.

❧

These notes that he left showed that his thinking was changing to that of the new world and was gaining all this knowledge from every source that he could find to develop his own view on how the world worked. It was at this stage where we see Newton believing in a world of particles, but also around this time, Newton's fascination with magic and alchemy would begin. This was all recorded in his notebook, which he entitled "*Certain Philosophical Questions*," which detailed the books Newton read and the ideas that he would have at the time.

❧

What wasn't recorded in his notebook was that it was during his period that his fascination with mathematical studies would begin. Again, he would turn to a book written by Descartes called La Geometrie where he would start to learn the basics of algebra and geometry before he would advance his reading. At this stage, Newton must have known that it all came naturally to him as he was reading difficult work with ease.

❧

It only took Newton about a year to become a master of the subject and had even started to look into his own line of critical mathematical thinking. It was at this stage that he would invent the basics of what would become calculus as well as binomial theorem where he would begin to work out how to calculate space, curvature, and everything between it.

❦

In 1665, the university had to be closed down due to the plague. Newton had only just received his bachelor's degree and without any normal training and gained his knowledge mostly from his books. Newton had become one of the greatest minds in the world, but he kept it to himself with his notebook. From here, we can see that Newton, in the two years that the university was closed, was working on his ideas. He took his ideas of circular motion and started applying them to the moon and the planets. He would lay the foundations for his own work on gravitation, but for now, he had kept all of these findings to himself.

❦

By 1669, Newton was already writing about his findings in a manuscript entitled "***On Analysis by Infinite Series***," which began to get him known within the mathematical community. This was the start of a point where we see Newton's reluctance to release his work to a broader scale, and a small number of his peers only knew his findings. At this stage, even though not a lot of people knew it, Newton was already one of the greatest mathematicians in the world.

❦

Upon his return to Cambridge after the plague, he was elected as a fellowship in Trinity College. At this time, Newton's studies were wide-ranging, and he was developing his understanding of optics, mathematics, and chemistry.

Also, Newton wrote a paper on fluxion and other branches of mathematics and handed into his professor Isaac Barrow. Barrow, with Newton's permission, shared it with an esteemed colleague John Collins.

Collins admired the paper, and it led Barrow to call Newton an "*unparalleled genius*," which just goes to show how highly regarded Newton was, and this isn't even listed with his most exceptional achievements.

In Cambridge, the role of '*Lucasian Professor of Mathematics*' was held by Barrow at that time who had previously taken Newton's works to London. Barrow stepped down from his role and recommended that Newton should succeed him though.

The role meant that Newton didn't have to tutor, but he was asked to deliver a series of lectures, and he chose his work in optics to provide the basis for his first lectures. This allowed him to develop an essay that he had written called "*Of Colours*" into a full book, which would be his first of his "*Opticks*" series. The lectures were hardly attended, but it still allowed Newton to develop some of his most important work.

❧ III ❧
THE START OF HIS GENIUS

"If I have seen further, it is by standing on the shoulders of Giants."

— SIR ISSAC NEWTON

THE BIRTH OF CALCULUS

❧

The story of calculus is not a simple one to tell, and Newton himself would probably be bored telling it as it wasn't one of his great passions and more of a means so that he could do his calculation on the universal laws of gravity. It would, however, lead him on a collision course to one of his famous rivalries.

❧

In its simplest terms, calculus is a way of calculating the rate of change and the areas, volumes and surface areas of complicated multi-dimensional shapes. It is for calculating the likes of a slope at any random point and focusing in on any point, which is called a derivative.

❧

Also, if you have the likes of a cube, then it's easy to calculate its volume and surface area, but with an odd shape, you need to use calculus as it will work out those figures by cutting the object down into infinitesimally small parts and adding them all together to get the correct figure. At its heart, calculus can do calculations with the types of shapes and curves that algebra and geometry can't cover.

※

To work all that out at the age of 24 was an incredible achievement, and calculus now forms an everyday part in our lives in more ways than many could imagine, playing an essential role in all the sciences, economics, and all branches of engineering. One of the marks of his genius was that he didn't actually set out of inventing calculus, but rather he saw it as a means when he reached a point in physics where he couldn't go any further until he solved the question of calculus.

※

Newton wanted to work out why the speed of a falling object increases over time and found out that there was no mathematical rule to work this out. Newton wanted to use this to not only work out this problem but also to work out planetary motion.

※

He desperately wanted to know why the planets moved in ellipses and not in a perfect circle as most planets when they are closer to the sun than at other moments. Comets especially have a strange orbit as they go close to the sun before being flung back out into far orbit. Newton broke it down into the orbit being a section on cones and was able to describe how the planets moved.

❈

To invent a completely new way of doing mathematics was a mark of Newton's genius, especially as he learned that new way of doing mathematics just so that he could work out planetary motion. In his Principia where Newton would explain calculus, he simply put it at the front so that he could explain everything that came after it.

❈

It was perhaps the fact that Newton saw calculus as a mean to work out planetary motion, or it could be the fact that Newton was always hesitant to publish his own work. But he wasn't the first one to publish a paper on calculus as Gottfried Leibniz first did it in 1684, long after Newton has started work on the subject. Newton was able to show that he had first invented it years before, but his lack of publication cost him dearly.

❈

It would start a very long debate at who actually invented calculus, and it turned out to be one of Newton's great rivalries, but now we understand that the two men had discovered it separately of each other. Newton, though, used his knowledge of calculus to advance his work on the planetary motion, which was his primary goal and eventually would make groundbreaking claims in the subject.

❈

Perhaps the most exceptional example of Newton's genius is that he wanted to solve the problem of gravitation and planetary motion and created a whole new branch of mathematics to do it. For what some people would be the crowning glory of all their life's work was just a means for Newton to answer the question that he really wanted answering.

The invention calculus by Newton has arguably made more of an impact than any other due to its use in so many things that we know of today. Calculus is used in the world around us every single day, and it's a branch of mathematics that didn't exist until around 350 years ago.

NEWTON INVENTS A NEW TELESCOPE

❦

I t is perhaps a mark of the genius of the man that he isn't more well-known as being the inventor of the reflecting telescope as it would change the way that the world was going to be able to look at the sky.

❦

The reflector telescope is one that uses mirrors, which may seem obvious now, but before Newton, the only way we could see long distances was with lenses that would bend the light. Newton's telescope would provide a much greater viewing distance and was also a lot easier to make. Not only that, they could be made much larger than the lens telescopes, which were called refractor telescopes.

❦

The genius was through the lack of distortion. Anyone who

looks through any glass knows that it can distort light; look through a glass at home, and you'll be able to see the problems caused by curved glass. If you're reading this through your glasses, you'll know how glass can alter light for the better. Those distortions meant that refractor lenses have huge limitations.

❧

The effect is called '***chromatic aberration***,' and it's where light is split into its various wavelengths just as it is in a prism, but to a much lesser extent. Newton rightly thought that it wouldn't be possible to get rid of the effect of chromatic aberration while still using a lens.

❧

As light doesn't pass through a mirror, this made Newton's telescope a lot more reliable. He experimented with numerous different metals and polishing techniques until he found the effect that he was looking for.

❧

The man didn't use this telescope to discover anything in respect of new planets or new moons. Instead, Newton used it to develop his understanding of gravitation and set out to mathematically prove the relationship that they had between each other.

❧

The actual telescope that Newton first made has survived and is in the care of the Royal Society that is situated in London.

Newton wasn't the first to think that mirrors would be a good idea for the inside of a telescope, but he was the first to make a working and usable reflective telescope.

❦

It wouldn't be for another 50 years that Newton's mirror would be improved upon by a man called John Hadley, who changed the shape of the mirror inside the telescope. Newton has just invented a device that would accelerate our understanding of the world. Not only that, but he would also use that device and the knowledge he had found to achieve some of the greatest findings ever known to man.

❦

It's a wonder what Newton would think today of the Hubble Telescope, which has been floating around Earth since 1990, taking some of the most amazing pictures of all-time with reflector mirrors that he first used in 1668 when he was just 25 years old.

❦

The telescopes of today have the technology that not even Newton would have dreamed of at the time, but the original mirror telescope was by him, and that's an incredible legacy to leave in itself.

HIS FAMOUS WORK ON LIGHT
AND COLOR

৩২৪৪৩

Optics and the nature of light were a richly debated topic since Kepler's "***Paralipomena***" in 1604, and Descartes moved that theory on to making light the center of his philosophy on nature and had argued that it consisted of motion, being transmitted through a materialistic medium. Newton accepted the theory of Descartes and started working on the different colors that could be found within the light.

৩২৪৪৩

Newton would show in 1666 that light was not simple, and it is instead a mixture of different colors with different properties that all come together to form a more complicated particle than the one that Descartes had thought.

৩২৪৪৩

From this, Newton was able to claim that the eye separates these colors, and his work would become the foundation of optics from thereon, and it was at this time that Newton invented the first reflecting telescope in 1668.

❧

Newton began experimenting with a prism in his early years, which involved a series of experiments that he had where he would see how light reacted with it. This, as we know now, resulted in splitting the light into all the colors of a rainbow - red, orange, yellow, green, blue, and violet.

❧

The thinking at the time was that light was just made up of light and darkness, and that the color, which came out of a prism, wasn't the splitting of light, but rather that color came from the prism itself. Robert Hooke thought that the red light, for example, as simple white light that had most of its darkness remaining, Newton disagreed.

❧

What Newton did to disprove this theory was to set up a prism near his window, which projected a large spectrum 22 feet away on his wall. To further prove that the prism was not coloring the light, he refracted the light back together. This was a demonstration that light was responsible for all the light that we see.

❧

At this time as well, there was a debate as to whether the

light was waves or whether it was a series of particles. No-one before had proved this, but through Newton's experiment, he was able to prove that light was indeed made from a series of particle and a series of colors that all combined to produce white light.

❦

It was Newton who first came up with the term 'color spectrum' and divided those colors into their separate parts. He was able to show that each color had its own angle of refraction, which can only be shown and calculated using a prism. Newton also had the thought that all objects are the color they are due to the beam of light that illuminates them. As a beam of refracted light never changed color, then the color must be the light reflected from an object, and not a property of an object. Therefore, if something is blue, then that is the color that it reflects.

❦

Again, Newton's revolutionary work on light was relatively unknown at the time. His work was delivered to the world through the recently established Royal Society of London, which had opened in 1660. The paper was all well received, just as his telescope was, but Newton had opened himself up to a world of judgment and critics.

❦

It was at this time that his rivalry with Robert Hooke would start as he was a prominent member of the Royal Society and had just published his own paper on light, which still claimed that light was made out of a series of waves. Hooke's name carried a lot of weight in the Royal Society, and many of their members agreed with Hooke's ideas and doubted the existence of Newton's spectrum as people had tried and failed to replicate Newton's experiment.

৩১৩

Prisms at the time were seen as something of a toy and not taken seriously as scientific instruments. They were merely seen as a form of entertainment, so when Newton tried to use one to prove a scientific theory, there was naturally some skepticism. The glass making abilities of the time weren't exceptional, and even the most well-made glass contained flaws and air bubbles. This was one of those times where Newton didn't help himself as he had concealed the details of his experiment.

৩১৩

Only four years after would Newton repeat the experiments and give more detail, so people could actually do the experiments for themselves. Newton couldn't handle the constant questioning of his experiments and withdrew his thoughts from public debate.

৩১৩

It was 11 years later in 1971 that the words of Newton's reflecting telescope had come, and the members wanted to see it for themselves. The response to the telescope was very positive, and it gave Newton the confidence to give them his paper on light and color. This was also the time when Hooke died, and Newton knew that his biggest critic in the field could no longer challenge him.

ॐ

In 1704, a year after Hooke's death, Newton published "***Opticks***," which detailed Newton's complete theory of light. At the start of the book, Newton finally placed a detailed account of how to reconstruct his experiments in fine detail. This led to many more successful reconstructions of this theory, and it seems like Newton had learned from these previous mistakes.

ॐ

There were parts of Newton's theory that have proved to be wrong though, as at this time, he still thought that particles of light created waves in the ether. The ether had originally been thought of as an elastic substance, which could permeate through all of space, including between the particles of matter. Therefore, it was believed that light was able to travel through this ether, and Newton thought that the wave properties of light were waves in the ether and not the light itself.

ॐ

The theory of ether, which was originally theorized by Aristotle, obviously turned out to be false. It wasn't until another genius who came along and eventually settled the debate at the end of the 19th century. So, was light a wave or a particle? Well, it turned out that everyone was right in a way. Light is both a wave and a particle as light is made of photons, which is a particle, and photons flow in a wave. The man who settled the debate was a certain Albert Einstein.

ॐ

Newton though in his lifetime may not have truly got to grips with the nature of light, but he did prove the properties of color and explained how we see objects in the color that we do. His work was revolutionary, and he was the first person ever to see color for what it really was.

NEWTON AND HIS RIVALS

❧

Isaac Newton was a man who had many battles throughout his life, and some of the greatest lifelong battles in scientific history involved the great man. A lifelong rivalry with Robert Hooke would drive Newton throughout the rest of his career. Hooke was one of the leaders of the Royal Society who had previously considered himself as a master in the field of optics. He wrote a condescending critique of Newton's work, and it's fair to say that Newton didn't take this very well at all.

❧

Newton was incensed by any criticism of his work, and Hooke hit him particularly hard and gave Newton the desire to humiliate Hooke as it did with anyone who crossed Newton's path in such a way. He was a man who wasn't able to rationalize any sort of critique, and he became so anxious about the discussion of his paper for which Newton would start to cut the social ties that he had made and withdraw himself into isolation.

❀❀❀

Newton would continue his work on color, and in 1675 when he produced a new paper explaining the properties of light, Hooke claimed that Newton stole the work from him. Newton became angry once again, but a series of letters between the two men eventually controlled the situation, but the two men clearly didn't like each other. It's possible that the genius of Newton might have threatened Hooke, and Newton was simply not able to handle someone questioning his work, whether it was valid or not.

❀❀❀

Another quarrel that Newton had was with a society in Belgium they also questioned his theory on colors. They questioned the validity of his experiments, and Newton would have a nervous breakdown after an exchange of letters. Newton fell into silence, and the death of his mother shortly followed. This was enough for Newton to withdraw from public life for six years and would only ever fleetingly enter into any conversation.

❀❀❀

Robert Hooke was a much-respected scientist but picked a battle with Newton that he shouldn't have done. Hooke will forever be remembered as the man who lost the battle with Newton, which is unfair in many ways, but Newton would get his wish to destroy his legacy. Hooke was a brilliant scientist and the first to use the term 'cell' to describe the individual units or a larger being. He also developed his theory of elasticity to explain the effects of spring. He pioneered the use of microscopes and was one of the first to determine that fossils were once living creatures.

❦

If Newton were alive today, he would have surely looked back on those moments and smiled; such was the malevolence of the man when it came to anyone who might cross him. Not only did Hooke dispute Newton's theory of light, but he also claimed that Newton couldn't have come up with his ideas from the Principia if it wasn't for him. Hooke didn't know when he was beaten; he would often come up with theories that lacked depth and that he couldn't prove. Hooke couldn't prove them, but Newton could.

❦

Hooke died in 1703; Newton wouldn't die until 1727, and Newton used those 24 years to discredit Hooke as much as he could, including getting rid of the only portrait that existed of Hooke in the Royal Society.

❦

It wasn't the only big rivalry that Newton would have though; he also had a big rival with another great mind in Gottfried Leibniz. It is now thought that both Leibniz and Newton discovered calculus independently. In their day, it was a debate that raged on. It was Newton's lack of publication that would cause his undoing, as Leibniz was the first to publish a paper on it when Newton had already been working on it.

In the end, Newton actually accused Leibniz himself of plagiarism, and the discussion was never resolved as there was never a definite dispute as both men had equal claim to working out calculus. Newton went as far as writing an anonymous report claiming that he was the inventor of calculus.

Leibniz would die 11 years before Newton, and again, he'd use the opportunity to discredit his rival further. In the end though, we can call his score draw as its common thinking now that no-one copied anyone. Newton at this stage was the President of the Royal Society and tried his best to use his influence on his own advantage. While Newton wasn't a man of many friends due to his genius, he had many followers who were more than willing to support him.

These weren't the only rivalries that Newton would have, and he'd have many more petty ones for the duration of his career, including one that he had with the astronomer John Flamsteed. Some of his work on comets has inspired Newton during his work on the Principia, and Newton did credit him. Flamsteed didn't think he received enough credit, so Newton instead decided to remove all credit from the second edition.

The rivalries show the dark side of Newton as he became enraged whenever anyone would try and dispute him. He was vindictive and malevolent whenever he was crossed and apparently was a man who couldn't help but hold a grudge, even after his rivals had died.

❦

Due to his genius and his status though, Newton didn't lose any of his battles. The one with Hooke being the most famous example, Hooke had a brilliant mind but is more remembered for losing his battle with Newton than anything he actually achieved. I'm sure, Newton would be as happy with that as he would be with his own achievements.

❧ IV ❧

THE MOST IMPORTANT
SCIENCE BOOK OF
ALL-TIME

"Gravity explains the motions of the planets, but it cannot explain who sets the planets in motion."

— SIR ISSAC NEWTON

THE PRINCIPIA

❧

I n 1969, Newton would continue his work on celestial mechanics, and he wanted to consider gravity and the effect that it had on the orbits of the planets. He took Kepler's laws on planetary motion and built on them. This would eventually lead to one of the greatest books ever written, which was the Philosophiæ Naturalis Principia Mathematica, which finally got published in 1687. Newton wrote the works in Latin, and the title translates to Mathematical Principles of Natural Philosophy. The title of the book is shortened to Principia, or Principles in English.

❧

Newton wrote the book in Latin even though English was then the dominant language of the time; Newton believed that anyone who couldn't speak Latin was worthy of being able to read his works. The book laid a platform for the rest of scientific history after it. The book stated Newton's famous laws of motion and the foundation of classical mechanics being Newton's laws of universal gravitation.

❧

It's one of the most important works ever created and would take Newton from being one of the most well-known person of his time to be regarded as one of the greatest geniuses to have ever lived. The Principia sparked a new age of scientific thought and started a great revolution in the world of physics.

❧

It wasn't immediately thought of as such though, and at the start, many people doubted the works and the significance of them. As time went on, the book continued to get an even higher status as its theory could not be unproven.

❧

The book sets out to solve the mystery of how massive objects move, and how the worlds around them are affected, and what affects them. He applies his thoughts and his hypo-thetical thoughts to both objects that are moving and to objects that aren't, this was done to work out which laws could be observed. Part of the genius of the works is its ability to explore the problems of motion being affected by external forces.

❧

The book showed many things including how the inverse square law of gravitation was proved and gave a prediction of the mass of the objects that were in our sky. He was able to work out why the Moon worked in patterns, which at the time were thought to be irregular and also theorized that the Earth was not actually a sphere and was actually oblate as the spinning of the axis makes the Earth bulge out in the middle, so, therefore, has the slight effect of a squashed ball.

❧

The book was also the first to explain the tides and showed how both the Sun and the Moon affect the oceans on Earth. It also explained how we have an equinox and explained how we have seasons, and it also explained the orbit of comets.

❧

The book went through and explained a whole series of events, which before the Principia were more of an educated guess. Here, Newton was able to show with workings and math how these events could occur. The Principia is divided into three books with each going on to explain different phenomena.

❧

In Book 1, which was subtitled '*On the Motion of Bodies*,' Newton would discuss motion in the absence of any resistance. It would help develop this theory of planetary motion and prove that large spherical objects attract other bodies outside of themselves as if all the mass was concentrated at the center. It allowed the inverse square law of gravitation to be applied to the real solar system to a great degree of accuracy.

❧

Book 2 wasn't as influential as either Book 1 or 3 as its claims weren't as far-reaching as it concerned resistance to velocity, the implication of resistance, and the different ways that resistance affects other objects. He also predicts the speed of sound and estimates it to be 1,088 feet per second, which works out to be around 714 miles an hour, which is not far off the correct figure of 767 miles per hour.

❧

Book 3 was titled as "*On the System of The World*," and it goes into many of the consequences on universal gravitation and builds upon the propositions of the previous books. This is where Newton would plot his theory about how the solar system worked and how the planets interacted with each other.

❧

The Principia also contained Newton's famous three laws of motion. Newton had initially started to develop these theories when he was 23 years old, but it wouldn't be until 23 years later that he would release them to the world. Newton's laws of motion may seem like an obvious part of life now, but they are fundamental principles that have stood the test of time.

❧

In Newton's first law, he states that an object will remain at rest or in a straight line unless an outside force is acted upon it. This is more or less the definition of inertia and the understanding that if there is no net force acting on an object, it will remain at a constant velocity or at rest if no force is acting upon it.

❧

The second law provided one of the most famous and simply the beautiful equations of all-time when Newton worked out that "***F=ma***," which can be applied to everything that we see around us. The "***F***" stands for force, and the simple way to work out force is to work out its ***mass (m)*** times its ***acceleration (a)***, which describes how objects change momentum. Newton used his calculus to confirm his findings, which would develop our understanding of objects in motion.

❧

Newton's third law is perhaps the most famous as it states thhhat for every action, there is an equal and opposite reaction. This explains a lot about nature as if object A exerts a force onto object B, then object B will exert the same level of force on object A. Just as if a bird flaps its wing, then it gets lift. If a baseball player uses energy to swing a bat, that same force is then applied when contact with the baseball is made.

❧

One of the astonishing aspects of the Principia is that it nearly didn't get published at all. In 1684, in a conversation with Christopher Wren and Edmund Halley, Robert Hooke claimed that he had worked out the calculations behind the laws of gravity and also the laws of planetary motion. Halley was not satisfied with Hooke's lack of evidence and couldn't work it out himself.

❧

Halley was one of the greatest geniuses of all-time and the

man whose comet bears his name but working out the laws of gravity was beyond him. After remembering Newton's works in the field of optics, Halley traveled to Cambridge to see if Newton would be able to solve the problem. It was during this time that Newton was living as a recluse, but he trusted Halley.

❧

When Halley asked Newton if he would be able to work on the solution, Halley would have been shocked at the answer as the great man told him that he had already worked out the laws of gravitation and how they affect the planets but couldn't find his workings to prove it. Unlike with Hooke, Halley believed Newton and eagerly awaited his work.

❧

He wasn't waiting for long, and a few short months after that conversation, Halley received correspondence from Newton and a 9-page paper, which was titled "***On the Motion of Bodies in an Orbit***," and thankfully the genius in Newton's head had moved from the dark and into the light.

❧

It was at this stage where Halley looked to have really inspired Newton as he became obsessed with his work in the field and wouldn't stop unless he absolutely had to. Newton's notebook on his chemical experiments went blank for about a year and a half as Newton obsessed over what would become the Principia.

❧

The Royal Society at that time agreed that this book was a masterpiece but did not have the funding to publish it as they had just spent off their budget publishing the "***History of Fishes***," which did not sell well. Instead, Halley took it upon himself to publish the work at his own financial cost. Robert Hooke made claims that the work was his, and when he did, Newton threatened to withdraw his third book. Halley, though, convinced Newton to release it as Hooke's claims could simply not be backed up.

꧁꧂

The book was then released in 1987. Halley would invent the diving bell, make the first weather maps, chart the path of comets, and work out the size of objects in our solar system, but perhaps his greatest contribution is motivating Newton to share his ideas with the world that otherwise might have been lost. Without Halley, we might have never known the incredible level of genius that Newton had.

꧁꧂

Not only was the Principia an incredible work of its day, but it went onto inspiring countless other scientists to build on Newton's work, including Albert Einstein who applied Newton's theory of gravity and used it to develop his own theory of space and time.

THE APPLE MYTH

❦

It's a nice and simple thought, isn't it? Newton, sitting there under the tree in the beautiful sunshine, minding his own business while he reads a book under the sun, then all of a sudden, an apple drops onto his head, and he suddenly has a moment of great thought, and all of the complexities of gravity are solved at once.

❦

It's quite simply one of the greatest anecdotes in the history of science, a field that has had no creative sparks. The story has seemingly been embellished by both Newton and anyone who has since told the story after him. There is an original manuscript, however, which does state that Newton devised the theory of gravity after watching an apple falling from a tree in his mother's garden, but there are no reports of it actually hitting him in the head.

❧

It was 1666 when it happened as it was the time when Newton had returned home from Cambridge after it had been closed due to the plague. At that time, he was obsessed with the moon and its relationship with the earth. He was trying to work out what the relationship was and speculated that gravity must extend to vast distances.

❧

After seeing the apple fall, he tried to prove mathematically that the force of gravity reduced at the inverse square of the distance. There is though, no personal account from Newton about the apple, just simply the manuscript that was written by the Royal Society.

❧

There is one account from William Stukeley who was an associate with Newton and used the fact that they were both from the same village as a means to start a conversation with the usually unfriendly man. Stukeley recounts a time when he was drinking under the shade of an apple tree with Newton, and he was told by him, in similar circumstances,

> *"why should that apple always descend perpendicularly to the ground? he thought to himself being occasioned by the fall of an apple."*

❧

It made him wonder why the apple should fall to the center of the Earth and not sideways and assumed correctly that there must be a drawing power in matter and that it must do so in proportion to its size.

***

There is another similar story from another man who recounts a conversation that he had with Newton, but both accounts came around 50 years after the event. Newton enjoyed telling the tale, especially as he got older. With it not being in any of his notebooks, you have to wonder whether it's actually true at all.

***

It definitely doesn't seem if the apple fell on his head at the very least, but whether it is true or not, for his mind to work in such a way as to theorize the theory of gravity in the first place was an incredible feat. Newton was a master of working out the world around him but seemingly also a master of telling a good tale.

NEWTON'S DARK OBSESSIONS

❧

I f Isaac Newton could come back to life today, if he had one thing that he could erase from his history, it would be alchemy. If someone tried to achieve what Newton did in the 21st century, then they'd be called crazy and perhaps just an idiot. With Newton though, you do have to accept his work in the time when it happened. A time before electric light, the telegraph and the industrial revolution had yet started.

❧

It wasn't just alchemy that Newton was interested in either; he also had an obsession with the bible and putting it in chronological order as well as trying to interpret it. It's a wonder what else Newton could have achieved if he didn't have these obsessions, but his lack of social engagements did leave him with a lot of spare time. A lot of time, however, he was more interested in these types of studies rather than his science.

☙❧

The study of chemistry wasn't very well-known in Newton's time, and we were still trying to work out how different elements reacted with each other. Harry Potter fans may know the term very well, but a lot of Newton's work in alchemy was in trying to find the philosopher's stone. In alchemy though, the philosopher's stone was a name given to a substance that could turn base metal into gold, a material which obviously never existed. It was also said that the material could be the key to achieving immortality.

☙❧

The thinking of the time though shows that this wasn't an idiotic endeavor for the time as The English Crown at the time would have given harsh penalties for anyone finding the stone as it would have devalued their own gold coins.

☙❧

Of the unpublished manuscripts that were found after Newton's death, it was found that around a third of them were related to alchemy. The reasons for the lack of publishing could have well been the fear of criticism that he might have predicted would come with such works. Not only that, but Newton was also reluctant to produce any works that he considered unfinished. We can see this from the 38-year gap that he had between discovering calculus and publishing it in his Principia.

❧❧❧

Further to his studies in alchemy, Newton was also obsessed with the bible. He wrote a manuscript in 1704 in which he detailed with attempts to get scientific information from the bible. He believed that through his studies of the bible and the temples that they described, he could uncover the secret wisdom that would give him the clues he needed to work out how nature worked.

❧❧❧

Again, in the time that he was living in, a lot of the views that he had were shared by philosophers of the time, and he certainly wasn't alone in having these kinds of thoughts. Newton specifically thought of himself as a man who was chosen by God to understand the scripture that was written in the bible.

❧❧❧

Even though Newton didn't write a whole book on the subject, he did have a lot of writing on the subject and even wrote a guide for prophetic interpretation where he detailed exactly what was required to correctly interpret the bible. He also tried to uncover a secret code in the bible. He didn't make any prophecies himself, instead saw himself as a messenger for the prophecies that were already written in the bible.

❧❧❧

Newton had a strong dislike for people who wanted to predict when the end of the world was going to happen, but

Newton himself made his own prediction, which itself wouldn't be discovered until 2003, where he predicted that the end of the world wouldn't happen before 2060. It wasn't a prediction as such, but an estimate from his bible studies. Around 300 years after he made that claim, Newton isn't far away from being proved right.

⚜

Also, with his studies of the bible, Newton wrote extensively on the chronology of the bible as well as of the ancient kingdoms. He tried to chronologically work out the history of civilization such as in Greece, Anatolia, Egypt, and the Levant. And a book, which was around 87,000 words long, was published shortly after his death. It's also noted in these books that Newton believed that the city of Atlantis once existed.

⚜

A lot of these beliefs may seem strange now, but if anything, it shows the way how Newton's mind works. At the time when he was living, so little was known about the world that you didn't know, which areas of study would lead anywhere, and which wouldn't. He applied the same methods of study to alchemy as he did to gravity.

⚜

It's easy, nowadays, to realize gravity as the most natural force that drives the entire universe, but imagine first learning about it and realizing that there was this invisible force that kept you rotating around the sun, and it was the same force that would pull the Moon around the Earth and the same force that would make an apple fall directly out of a tree.

❧

It is equally astonishing to realize that the white light, which everyone was used to, actually splits into a variety of different colors. Newton and others were fundamentally changing the way that people viewed the world. At the time, believing in the existence of gravity would have been as crazy as believing that there was a material that could turn a base metal into gold.

❧

While some of Newton's beliefs seem strange now, they weren't seen so in his day. It's speculated that Newton was a member of a few secret societies that would discuss such things. In the future, it might be discovered that there is something that changes our view on the world and makes out today's thoughts seem foolish. While Newton was clearly a very odd man at times, he wasn't crazy for having such thoughts on alchemy, bible scripture, and chronology.

V

NEWTON THE MAN AND HIS LATER LIFE

"What we know is a drop, what we don't know is an ocean."

— SIR ISAAC NEWTON

NEWTON THE HANGMAN

❧

From 1696 until he died in 1727, Newton was the Warden and then the Master of the Royal Mint, which was a curious change of career after dedicating his life to mathematics and science. He applied his scientific methods into his role and set about to change the culture of counterfeiting coins, which was so prevalent in the day.

❧

Before Newton, this was meant to be seen as a comfortable job for men of high esteem where they would relax and delegate all responsibilities of work. Newton though, just like when he was asked to be a farmer, couldn't contemplate a life where his mind wasn't always working on complex tasks.

❧

Newton wanted to make all coins as accurate as possible and

tried to make them hard to copy so that they would be reliable. Under Newton, the Royal Mint's coins became the most respected in the world. He would bring the currency into a new age of craftsmanship.

❧

Naturally, Newton refused to accept any criticism of his coins as was proved in the Trial of the Pyx in 1710 where an independent jury weighed and assessed the Royal Mint coins to confirm their worthiness. At this trial, it was reported that they thought the represented gold coins had some issues with their weight. Of course, this enraged Newton as someone had dared to question his work. Newton believed that he was in the right and eventually questioned the trial plate that was used to assess the coins. After a test, Newton was proved right, and the plate was changed, and Newton had proved his point.

❧

At the time, Newton had estimated that around 20% of all coins in circulation were counterfeit when the Mint had a **_great coinage_** in 1696. Prosecuting the criminals was difficult due to the lack of proof, and they had been left to make these coins unchallenged for many years, but that was to change with Newton. One of the methods he took was to disguise himself and go to the bars and taverns, collecting most of the evidence himself.

❧

He conducted over 100 cross-examinations of various involved people and ended up prosecuting 28 people for the crime. It was a landmark time for the Royal Mint and showed how people, who try to create fake, would be brought to justice; it was defined as an act of treason, so the punishments were severe.

❖

One such famous criminal was a man called William Chaloner who underestimated Newton's ability to prove that he was a counterfeiter. Chaloner was a man who was trained as a nail maker but would use that training to make coins instead, coining 30,000 guineas. It enabled Chaloner to have great wealth to pose as a gentleman.

❖

He claimed that Newton was incompetent and had blamed the Mint for the fake coin issue and denied any responsibility. As you can imagine with his competence being questioned, Newton wasn't too happy about the accusation and placed even more effort into solving the problem.

❖

Newton soon had enough evidence so that Chaloner could be imprisoned, but then Chaloner used his wealth to bribe the star witness out of testifying. Chaloner was released, and he accused Newton of jailing an innocent man. If he hadn't already done enough to enrage Newton, he had certainly done it now as Newton decided to try and bring him to justice by any means required.

❖

This included bribing crooks for any leads, making threats, and pressuring Chaloner's associates. After two years of pursuing the case, Newton once again had him back in prison. This time though, the case went to a full trial. In the battle of wits and cunning, Newton had won. Chaloner ended up paying a big price for this with his life, as he was hanged for his crimes.

This was all before Newton was even a Master of the Mint as at this stage, he was still just a warden of it. The Mastership was offered to him, which he took up in Christmas Day in 1699. Newton took the job seriously and resigned from his position in Cambridge two years later.

In 1717, Newton wrote a report about gold and silver coins, and as a result, a Royal proclamation was decreed, which forbade the exchange of gold guineas for more than 21 silver shillings. This led to a shortage of silver, which moved Britain out of the silver standard and into the age of gold standard.

Under Newton's watch, the currency in the country had developed and become a lot more respected and valued. It's a part of Newton's career that receives very little attention but was a hugely important role for him. It was during his tenure at the mint that he would get knighted by Queen Anne.

The ruthless pursuit of Chaloner also shows off Newton's vindictive side where he wanted to win at all costs. He knew that the fate of the men he would catch would be to hang, but he thought that it was a price worth paying for his crimes. Newton was very good at winning his rivalries that he had, with this particular rival, he could directly claim responsibility for his death.

⚜

It's a wonder what Newton would think today of the new notes that have been issued in Britain that are plastic and near-impossible to counterfeit, I'm sure he would have been impressed by the technology. He'd also be happy that his face was on the old one-pound note in Britain as well as a new 50 pence coin being made in his honor recently.

⚜

While Newton will be rightly remembered for much greater works, he saved the economy, millions of pounds in today's money, by clamping down on counterfeit coins.

NEWTON'S WEIRD AND
WONDERFUL PERSONALITY

❧

Sir Isaac Newton was a very complicated soul who could be very difficult at times. Whether that was the result of such a difficult childhood is unclear but having a father who died before you were born and a mother who would abandon you while you were three aren't exactly going to help any young boy find his way in life.

❧

Newton never married and was never one who would sustain any kind of friendship with anyone. It's hard to judge Newton's mindset, but it doesn't appear that he was a mean person for the sake of it, but one that had little interest in people. He never had any interest in traveling and could always be found in either his hometown, Cambridge or London.

❧

Perhaps, Newton's most flawed characteristic is his anger, especially if anyone was to question his work. The man would become irrationally angry, and it would lead to a moment where he could be vindictive, but also petty as well.

❀

Despite his abrasiveness, Newton had an astonishing career and was served well by his genius. His personal vendettas against Robert Hooke and Gottfried Leibniz showed his ruthlessness, and while his angst against Hooke especially was justified in many regards, it was a vendetta that Newton couldn't let go, and he persisted this trying to discredit them long after they had passed away.

❀

This vindictive side of Newton was further shown when he was the Master of the Royal Mint when, in those days, it was meant to be an easy job, which required very little work. Instead, Newton went after the counterfeiters that had poured so much fake currency into England that Newton believed its level to be 20%. The crime of counterfeiting was one of treason, and Newton had many such people hanged for their crimes.

❀

It would be interesting to see what a psychiatrist would make of Newton today as Newton shows all the classic signs of Asperger's syndrome. He was socially impaired with poor nonverbal communication, a lack of empathy as well as not developing friends. He had a lack of interest in communicating with others and had an obsessive impulse to adhere to routine.

❦

Newton, by all accounts, would most likely have died a virgin as sexual pursuits never interested him, and he was always one who was very wary of any exchange of bodily fluid. Newton was a man who was obsessed by his work, and there was never any account of him keeping any romantic female company. The reasons behind that were unclear as we do not know Newton's views on sexual matters. It could have been his fear of exchanging bodily fluids, or it could have just been his extreme anxiety. Voltaire, who was an associate of Newton's, claimed that he had

"neither passion nor weakness"

and that he

"never went near any woman."

❦

It does also appear that Newton suffered from depression at a time, and his complicated personality led him to withdraw from life for long periods of time. Due to the upbringing he had, Newton can hardly be blamed for not being socially rounded. By all accounts, his father was a quite extravagant man who could be the life of the party.

❦

Newton's father, though, was obviously never there to look

after his son and never there to help him develop into a more rounded social man. Newton would have had issues that would come from his mother as well, especially with the abandonment that he would have felt. This might explain Newton's strong reaction whenever he received criticism. The rejection that he would have felt from his mother could well have fed into a fear of rejection with his works.

That fear of rejection would mean that the laws of gravity may have never been discovered had Halley not persuaded Newton to place his thoughts into his book. Not only did Newton make the discoveries, but it has been said that people had to discover Newton's discoveries as he had kept most of them to himself for a time. Newton was a man who was full of fear, but also had an arrogance about himself and his mind; an arrogance that many people would say was justified.

He was to suffer at least two mental breakdowns with the first one being in 1678. This was the first time Newton had put his work out there to a wider audience, and the critic from Robert Hooke was too much to handle for Newton. The next one, in 1693, was very severe as the mental breakdown was made worse by his insomnia, loss of appetite, and also having paranoid delusions. This led Newton sending strange correspondence to family and friends, which he would later apologize for when he recovered his health. It was thought that Newton, at this time, would have been suffering from the ill effects of mercury on account of his alchemy projects, but he didn't show signs of any other symptoms.

❧

Like many great minds, Newton was deeply troubled, but thankful circumstances meant that it didn't affect his work too much, but sadly, it doesn't ever seem to appear that Newton was happy anymore. Maybe that was the price for his genius, a mind so great that it couldn't reason with mere mortals. Newton had a high opinion of himself, and with good reason.

❧

It is a shame in many ways that he didn't marry or have children and never settled down. His loss was the world's gain though. He was a depressive man who was very insular and vindictive when he was challenged. He's the type of man that you can imagine leaving comments on YouTube and Twitter, correcting people and putting them right if he lived in the modern day. He was a very interesting character, but if you were ever asked to play that game where you can pick any dinner guests alive or dead, then you probably shouldn't be putting Newton in your list if you wanted a good time.

HIS FINAL YEARS

❧

Isaac Newton lived to the old age of 84 and died in 1726. After a life of such fierce passion and rivalry, he died in his sleep after previously suffering from pain in his abdomen. Newton would close his eyes on that night and never regain consciousness. In the last few years of his life, Newton had increasing levels of issues with his digestive system and made efforts to try and change his diet.

❧

He died unmarried with a few friends around him. He never left a will as most of his estate had already been left to members of his family before his death. Following on from his death, his hair was tested, and it came back positive for mercury, and it was thought that mercury poisoning could have played an important role in his health and also his increased eccentric behavior in later life.

ॐ

Sir Isaac Newton was buried at Westminster Abbey along with some of the greatest minds of all-time and is buried at the same part of the church as the likes of Charles Darwin and Robert Stephenson. Recently, Stephen Hawking was interred next to the grave of Newton, and they both had been Lucasian Professor of Mathematics at Cambridge University, Newton being only the second person ever to hold the position and Hawking being the 17th.

ॐ

Newton lived a long and colorful life and continued to work for as long as he could as he still held the posts of Lucasian Professor of Mathematics and the President of the Royal Society until his death.

NEWTON'S LEGACY

❦

When it comes to the greatest genius of all time, Sir Isaac Newton is in that conversation as possibly the best mind to have ever lived. The man achieved feats that would have been deemed unthinkable at the time.

❦

Newton didn't achieve his feat in one field, but numerous fields. He was a genius in mathematics, optics, as well as physics, and his achievements in each of these categories would have had him listed as a genius on their own.

❦

His legacy is one of a man who left the world in a completely different way that he found it. By the time of his passing, he had delivered to the world a lesson in many of the fundamental aspects of science that we take for granted today. The 16th century was devoid of invention and scientific thought; the world was still in the dark ages about what the words really meant.

***

Newton was born midway through the 17th century, and that's when the world that we know today started forming into life. A collection of scientists and philosophers would develop theories and make discoveries that would bring the world to life. Out of those great minds, Newton's shone above them all.

***

In the 18th century, in which he died, we start to see the introduction of electrical energy, a much greater understanding of the universe, and the start of the industrial revolution, a revolution that wouldn't have been able to happen so quickly if it wasn't for Newton's findings.

***

Newton's first real mark on the world was with his reflecting telescope. His idea of using mirrors for a telescope instead of lenses was revolutionary as is still the fundamental principle that is being used in telescopes today. There are telescopes panned which will have new mirrors that are over 100 feet across, and a new telescope is being launched into space called the James Webb Space Telescope, which will once again be based on mirrors as was first achieved by Newton.

***

He would go on to make discoveries within the light that no other had managed to do before. One thing that Newton could never quite work out was the nature of how light moved and still held onto the same belief that Aristotle had about it moving through an ether. While Newton may not have been able to work that out, his work with color was groundbreaking as he was the first one ever to prove that light splits into different colors and that the color what we see of objects are simply just the light that reflects off of them.

⚜

Further into his career, Newton would release the work, which he'll forever be associated with, the Principia. It firstly showed how he invented calculus, a completely different way of looking at the world through the medium of mathematics. Newton was many things, including one of the greatest mathematicians to have ever lived. He had a natural gift for the subject, but a gift that didn't even rank highly on what he wanted to achieve.

⚜

The discovery of gravity was what Newton will be remembered for though, but while you would say it is his greatest achievements, but sometimes it overshadows all the other incredible feats that he made. To be the first person in the world to work out how the planets orbited around the sun was a finding that almost defies belief. To have such a unique mind to work out the relationship between mass and velocity is what makes Newton one of the greatest geniuses of all-time.

❦

His laws of motion are still being used today as fundamental laws, which changed the way that people thought about motion, and the calculations that were made from that became the basic calculations that are made in all engineering projects today. Most of Newton's theories have remarkably stood the test of time, with only Einstein making slight adjustments to Newton's laws of gravity when he discovered that mass bends space-time, a discovery that Newton simply couldn't have made with the resources he had available to him.

❦

Newton was apparently a troubled man and probably not too pleasant to have a conversation with unless he had your respect. His traumatic childhood in no doubt contributed to that, and also the way that he was unable to deal with criticism was probably due to the rejection from his mother. He had some famous battles with some of the greatest minds of the time such as Leibniz and Hooke; history will record those as a win for Newton as it's his name that is remembered much more fondly.

❦

It is perhaps a mark of Newton's genius that there's a remarkable lack of people who wouldn't ever know that he became the Master of the Mint and was responsible for the hanging of many men. He showed the same dedication to his pursuit of counterfeiters as he did to alchemy and gravity. He was relentlessly obsessed with his work, and the results of that obsession speak for themselves.

Sir Isaac Newton's legacy is one of the greatest and perhaps the most celebrated geniuses to have ever existed. If you look at any poll of the greatest minds of all-time, Newton will be up there. That level of respect is more than deserved. In a time when there was an explosion of scientific thought, his mind stood higher than everyone else.

Newton once said,

> *"If I have seen further, it is by standing on the shoulders of giants,"*

but ironically, Newton himself became the biggest giant of them all. Everyone after him stood on his shoulders, Einstein, who is a competitor with Newton when it comes to the greatest minds ever, couldn't have achieved what he did without Newton's discoveries.

For most people, they will remember Newton solely for his work on gravity, but he deserves to be remembered for much more than that. His groundbreaking works in color, light, telescopes, calculus, mathematics, and physics make him quite simply one of the most important humans to have ever lived.

❧ VI ☙
THE STRENGTHS AND WEAKNESSES OF SIR ISAAC NEWTON

❧❦❧

Newton was a genius, and in many ways, it's hard to compare yourself to someone who had a mind far above what almost every other person has. Newton, though, recognized his gift and worked hard to use his intelligence to his advantage.

❧❦❧

We see many people with great minds or many sportsmen with great talent not making the full use of their potential. Newton made full use of his as he worked from morning to night on his theories. He showed what could be achieved if you set your mind to it.

❧❦❧

He used his intelligence to surround himself with people who would support him and defend him whenever he needed to. Newton also went about some of his likes in a way that he would try and crush his rivals, just like a ruthless businessman.

⚜

He also showed the values of not giving up. Many of his theories took a great deal of work and patience. You can imagine that there were many times when Newton would have been very frustrated with not being able to complete a theory. Newton never gave up though and persisted until he found his answers.

⚜

Regarding his weaknesses, Newton clearly didn't have any when it came to his application of science as he never let anything get in the way of his work. There are no reports of Newton ever being a happy man though. He was reluctant to have any social engagements and didn't ever find a woman and didn't ever come close to having children.

⚜

It was clear that Newton would let the tragedy that he suffered in early life and the abandonment that he felt affect him for the rest of his life, although you can hardly blame him. This wasn't a time when such thought could be discussed, but Newton was undoubtedly a complicated character to deal with.

⚜

The vindictive side of his personality was probably his worst trait as he relentlessly went after his rivals until he won. He could be nasty in many ways, but seemingly only cared about his work.

⚜

For the world that he left around him though, it didn't care what type of personality he had, but what he achieved made us change our view of the world.

❧ VII ❧
HOW CAN WE USE NEWTON'S STRENGTHS IN OUR LIVES?

❧

Newton was a man who showed the value of hard work. While we may not be able to have the mind that Newton did, he was still able to show that you can get to your full potential if you give yourself the right amount of application.

❧

Newton was a meticulous worker who devoted himself to his studies. He knew that he was building upon the work that had previously been completed by others, and he didn't shy away from the fact that people greatly appreciated the work of the people who came before him. Newton also showed the value of patience, which can be a crucial part of everyone's lives. We all reach those moments when we feel like giving up on a task, but Newton was the type who would never give up on a task until he received an answer.

⚜

Newton's greatest strength was undoubtedly his genius mind, and it's clear to everyone that it's a strength that only a very small few ever had. We all have the potential to make the most of what we have though, and that's the inspiration we can take from Newton. He took the cards he was given by God and used them to make as much out of his life as he could. While most others aren't gifted with the same genius; if we have the same level of application, then we too can make the most out of our potential. While you may not be able to discover new ways of looking at the world, but you'll still be able to achieve greatness.

⚜

Newton was an inspirational man in many ways, but if it weren't for his hard work, dedication, and patience, then we wouldn't have been able to achieve anything. While he may have been the greatest mind of his time, he also showed the greatest dedication to his work.

❧ VIII ☙

THE BEST BOOKS ON
ISAAC NEWTON

❧❧❧

The Principia: The Authoritative Translation: Mathematical Principles of Natural Philosophy - Isaac Newton and I. Bernard Cohen - The Principia isn't an easy read, but it is one of the best books of all-time. This is an excellent translation that will guide you through Newton's greatest work.

❧❧❧

Isaac Newton: The Last Sorcerer – Michael White – This is a fascinating look into Newton's obsession with alchemy and the influence it had on his life.

❧❧❧

Priest of Nature: The Religious Worlds of Isaac Newton – Rob Iliffe – A deeper look into a deeply religious man and his sometimes-extreme views.

ALBERT EINSTEIN

INTRODUCTION

❦

Most readers do not realize that Albert Einstein was a man of many faces. He overcame many obstacles in his childhood to become the person he was as an adult.

❦

That being said, I will let you be the judge as to the person he became as an adult. We all know he was a pure genius, but what kind of person was he outside of his world of genius?

❦

Even though he was a man that not only wanted peace in the world, he still created **a *formula/theory*** *that would build* **a bomb** *capable of killing thousands at a time.*

❧❧

His formulas have led us to such modern technology that we might not have today in the palm of our hands and changed the way we live our lives.

❧❧

Through *Albert's breakthroughs in physics*, he has made it possible for us to have items such as televisions, global positioning satellites, computers, CD players and millions of more things too numerous to mention that are at the touch of our fingertips every day.

❧❧

He has helped bring forth inventions that have simplified our lives more than anyone ever dreamed of at the time of his discoveries.

❧❧

As far as the way he lived his life, well, read on, and you be the judge to see if you think he had a full and happy life. Einstein's story awaits you on the pages ahead.

❧ II ❧

A GENIUS SHOWS UP
WITH A DEFORMED HEAD

"Few are those who see with their own eyes and feel with their own hearts."
ALBERT EINSTEIN

❧

The place was Württemberg, Germany and the date, **March 14, 1879**, at 11:30 AM when a baby boy was born to *a feather-bed salesman and his wife*. No one at that time had any inkling that one day this baby would stand the world on its head. The child's birth weight cannot be found recorded anywhere in history and for sure not on his birth certificate. That more than likely could be attributed to the fact that few had a means of weighing a newborn at a home birth back in those days.

❧

This town today has a population of about 120,000 and a small plaque is all that is left to commemorate where Albert Einstein's home stood when he was born. The house was destroyed during World War II.

❦

Albert's mother, Pauline, noticed after his birth that his *head seemed unusually large*. She could not help but worry about it and could not get it out of her mind. She noticed that it was oddly shaped, mostly on the back side.

❦

One thing that upset both parents was the way he was developing intellectually. His speech was so delayed that he did not begin to speak until he was 3 or 4 years old and not fluency until he was nine. Even with lack of speech development, he still managed to be one of the top students at the elementary school he attended.

❦

Looking back, researchers at Oxford and Cambridge Universities feel that Albert showed signs of *one form of Autism called Asperger's Syndrome*.

❦

Many who have been diagnosed with Asperger's are thought to be eccentric. They seem to lack social skills and seem to be obsessed with topics that are complex, and they may have issues communicating with others.

Researchers think that Einstein, from a young age, exhibited signs of Asperger's.

Researchers think that Einstein, from a young age, exhibited signs of Asperger's.

There were signs when he was a child. For instance, the way he would pull away from others and be such a loner and how *he would repeat sentences over and over*. Albert would keep up the repeating until he turned seven years old. The Asperger's even reared its ugly head later when he became a professor and gave lectures. So much was going on in his mind at once that it seemed to confuse what would come out of his mouth.

Researchers also say that his passion, standing up for what he thought was justice, falling in love; all are exactly compatible with Asperger's. But, what most individuals that suffer from Asperger's can't do is small talk. So, if he found the right woman that could mingle well in social situations, he would not have to chit chat at social events as she could do that for him.

When Einstein grew older he made intimate friends, he spoke out on about every political issue you could imagine, and he had numerous affairs even though he was married.

His mother's maiden name was Pauline Koch, and she made

sure the families household ran like a well-oiled machine. Albert was joined two years later by a sister who was named Maria.

৩৶৶

When Albert was five years old, his uncle gave him **a compass**, and it immersed him in the fact that *an invisible force could make that needle move*! That compass alone at age five was what would start him on his journey through life with anything that had to do with invisible forces.

THE ODD SHAPED HEAD STARTS
TO READ

᪉

H is father later achieved with moderate success the oversight of *an electrochemical factory* after the family moved to Munich.

᪉

While living in Munich, Albert turned twelve years old; he happened upon a book called **"geometry,"** which he lived with his nose inside of day and night. He often referred to it as his "sacred geometry book."

᪉

At twelve, he also became very religious. He wrote many songs of praise about God and would chant gospel songs on his way to school. It started changing after he began reading some of the science books that were in direct contradiction to his beliefs about religion thus far.

◈

While attending school at the Luitpold Gymnasium, Albert felt out of place and bullied, as we call it today, most of the time because of the education system used by the Prussians that meant to hold back those who seemed to be creative and original in their thinking. One teacher at the Gymnasium went so far as to tell Albert he would never turn out to be anything.

◈

Albert *learned to play the violin*, and he was quite good at it, and he enjoyed playing it. He had never become interested in music until he had heard the music of Mozart and then he was hooked. He was a fan of Beethoven and Bach, but **Mozart** was by far his favorite. It was said that playing his violin or the piano helped him when he was working on his theories. *He could play a few notes on the piano or the violin, and then he would jot down notes on some theory.*

◈

But, thank goodness for a young medical student by the name of **Max Talmud** who often came to their home to dine with his family. Talmud started out by working as a sort of informal tutor to Einstein on a higher level of learning in philosophy, science, and mathematics. Talmud made Albert's fascination with all things regarding physics even more heightened.

◈

One very pivotal moment in Einstein's life was at age 16

when he read **"Popular Books on Physical Science,"** wherein the author himself talked about being able to imagine riding so fast that you could keep up with what was traveling inside a wire. Einstein could barely think of anything else for the next ten years. What would it be like if you could run beside a beam of light at the same speed? If perhaps you could do this, it would seem as the beam of light was standing still.

❦

During this time, Albert had written his very first of many **"scientific papers,"** (The Investigation of the State of Aether (what we know as ether) in Magnetic Fields").

❦

Albert's father never seemed to be able to get anywhere with any of his businesses. In 1894, his company lost a contract they needed to stay operational, so *the family had to move to Milan* for his father to work with one of his relatives.

❦

They left Albert by himself at a boardinghouse in Munich and told him to finish his education there. Albert was miserable and despondent, so alone and utterly repelled by the idea that he would be required to serve military duty when he turned 16.

❦

Einstein ran as fast as he could six months later and wound up on his Mother and Father's doorsteps. They were shocked to see him and anxiousabout what he would be facing as **a**

draft dodger, school dropout, and he had no employable skills. What in the world were they going to do with him?

⚜

But, desperate times do call for desperate thinking and drastic measures and out of that can come some great answers. It was with great luck that *Albert was able to apply to The Swiss Federal Institute of Technology*, even if he did not have a high school diploma, if he could pass the tough entrance exams.

⚜

Albert excelled in physics and math, but *he failed at chemistry, French, and biology*. Because Albert had such exceptional math scores, they allowed him to go ahead and enter the school on one condition, and that was if he first finished formal school.

⚜

Albert attended a "special" (a progressive and free-thinking) high school **in Aarau, Switzerland** which was run by **Jost Winteler** and graduated within three semesters in 1896. While attending there, he lived with Jost Winteler and his family.

⚜

Albert had begun what would become a lifelong friendship with the entire Winteler family while living with them during the three semesters. *Albert's first love was Winteler's daughter,* **Marie**. Albert's baby sister, Maja, would later marry the son of Winteler, Paul.

❦

At times when Einstein looked back on his life, he would have to say that his happiest years were those spent in Zurich. He met many loyal friends and lots of students. It was in Zurich that he met his first wife **Mileva Maric** who was a physics student from Serbia.

❦

Mileva was the only woman pursuing a physics major in Zurich while Einstein was attending the college. It was during the second semester that the two became interested in each other.

❦

Einstein's mother was against Mileva from the beginning. It did not keep Albert from romancing Maric while the whole time his Mother and Dad were not happy with what looked like "soon to be a wedding." His Mother did not like her because she had a Serbian heritage and her religion was Eastern Orthodox Christian.

❦

After Einstein graduated from Polytechnic, he took a job away from Zurich, while Mileva had to stay there in Zurich. Mileva was not doing well with her grades as she failed her final exams twice.

❦

Albert had convinced Mileva to meet him at *Lake Como* for a romantic weekend vacation. It was a well-known resort that had beautiful scenery of snow-capped mountains. It was this weekend when Mileva conceived their illegitimate child.

❦

Einstein would come to see Mileva every Sunday to visit. It was when Albert came during one of those visits that Mileva told him she was pregnant.

❦

Meliva quit school because she had failed her physics exams two times. She was so depressed that she went to Hungary to her parent's home where she had to tell them about both of her failures, facing them all by herself. In the beginning, her father said she could not marry Albert.

❦

Then came the winter of 1902, Mileva delivered the baby girl that they had named **Lieserl**. Mileva had a difficult time during the birth process, and Einstein was not there to support her. He found out about the baby through a letter from Mileva's father. Einstein was so excited about his new child that he wrote back full of questions. He wanted to know if she cried adequately, if she was healthy, who she looks like. He was so happy and so in love with the baby and had not even seen her yet.

❦

In the letters between Mileva and Einstein before Lieserl's birth, Einstein had told her to nurse the child instead of giving the baby cow's milk. He was afraid the cow's milk would make the baby stupid.

❧

It was in 1902 that Einstein was lower than the lowest in his life. He was not able to marry Maric because he didn't have a job, and therefore, could not support a family. His father's business went belly up again. Einstein was unemployed and desperate and finally decided to take one of the lowliest jobs you could get, and that was tutoring children. As luck would have it, he even got fired from tutoring.

❧

It was about that time that Albert's father took to *his deathbed*, and right before he passed away, he told Albert he had *his blessing to marry Maric/Mileva*.

In the late part of 1902, a bit of good fortune came Albert's way. A friend of his father, Marcel Grossman got him on at the Swiss Patent Office in Bern as a clerk of sorts.

❧

For some reason, despite all his interests, Albert never traveled to meet his new infant. No one knows why. It is where the story about the baby starts to get darker. Maric moved back to Zurich to wait for Albert to get the job at the patent office so they could get married, but she did not bring Lieserl with her. *Albert and Maric married in January 1903*, and they moved into an apartment.

❧

It did not keep Albert from romancing Maric while the whole time his Mother and Dad were not happy with what looked like that might soon be a wedding. His mother did not like her because she had a Serbian heritage and her religion was Eastern Orthodox Christian.

❧

More curious than ever is in all the research that has been conducted and all the letters back and forth between Albert and his wife there seems to be no common thread running through the stories of the time. August 1903, Maric found out that *Lieserl who was about 18 months was sick with scarlet fever*. She went to see the baby. In September, Einstein wrote to Maric who was still there with the baby. He was worried about their little girl. He asked her if Lieserl was registered. He told Maric they had to be careful so that it did not cause the child problems in her future.

❧

No one knows if the tiny child died of Scarlett Fever or maybe they gave her up for adoption.

❧

Giving her up for adoption seems far-fetched when you realize the streak of stubbornness Einstein wore all the time. He never gave in or gave up on what lay before him. In that period, it was common that children died of scarlet fever and as young as the baby would have been, it could very well be a definite possibility. It shall forever remain a secret to all unless somewhere a letter that has never seen the light of day comes forth that reveals **this secret**.

✦

While Maric was away, she realized she was pregnant again, and this time it was a boy, **Hans Albert Einstein**.

✦

Nothing can be found as to what happened to Lieserl. So many reporters, journalists, and researchers have searched for this missing child of such a brilliant man, and yet, she has disappeared without a trace.

✦

Albert did not seem to be a perfect husband, but it must be said that he did try in the beginning. He was just so into his work and with all of his attention on his work that it was like he did not even know his family was there. It was like he was not "in the present" with the rest of his family.

✦

Albert dug into his work even more. Mileva sank into depression. (One must wonder if Mileva did not suffer from postpartum depression as it had not yet been discovered at that time.) Their house, seen by one visitor, was an absolute mess. It is said that Albert did try to help, but he couldn't get into it. He would pick the baby up and push the stroller around and be writing equations on his notepad all at the same time; he hardly realized he was a father and had father duties.

❧

July 28, 1910, **Eduard Einstein** was born. Things seemed to improve for them for a while, but it didn't last. Mileva was still depressed, or maybe she just had post-partum depression again, who can tell by the writings that have been found and she was getting jealous of all those women that Einstein openly flirted with in front of her or those he bragged about to her face.

❧

In 1910, Albert's mom moved to live with her sister Fanny and the rest of Fanny's family to Berlin after Albert's father had passed away. Now bear in mind, **Elsa** (*Albert's second wife*) was Fanny's daughter. Pauline took a job as a housekeeper in 1911.

❧

She eventually moved to live with her brother, Jacob Koch, in Zurich in 1914. During the days of World War I, Pauline become ill with cancer.

❧

Einstein and his little family moved to Prague in 1911, where he was going to teach at the university. Meliva did not like living in the city.

❧

One year later, Albert had an offer from Zurich, and they moved there. Mileva was so happy to be in Zurich. They were there only a couple of years though, because *in 1914, Einstein was offered a position at the University of Berlin* and again they moved.

❧

Mileva hated moving there and was very unhappy. Albert's cousin, Elsa, would be close at hand for him. Mileva was very jealous of her.

❧

And Mileva's suspicions were spot on; Albert started dating Elsa. It started the end of Mileva's and Albert's marriage.

❧

The marriage kept deteriorating, and they attempted to glue it back together for the children's sake. Einstein, as a pacifist, sat down and wrote a "list of conditions" that his wife had to accept when he got home if they were going to stay married. The following is the verbatim list he gave her.

❧

CONDITIONS:

- *You will make sure:*
- *That my clothes and laundry are kept in good order;*
- *That I will receive my three meals regularly in my room;*
- *That my bedroom and study are kept neat, and especially that my desk is left for my use only.*
- *You will renounce all personal relations with me insofar as they are not completely necessary for social reasons. Specifically, You will forego:*

- *My sitting at home with you;*
- *My going out or traveling with you.*
- *You will obey the following points in your relations with me:*
- *You will not expect any intimacy from me, nor will you reproach me in any way;*
- *You will stop talking to me if I request it;*
- *You will leave my bedroom or study immediately without protest if I request it.*
- *You will undertake not to belittle me in front of our children, either through words or behavior.*

This list is chilling and to think that he would make such demands of the womanhe had been so in love with in his early years makes it hard to comprehend. What had happened to them as a couple?

I do realize that it is not fair to judge someone this far back in time and not know all the details of both sides, but it is hard to read this and not feel chills down your spine.

Einstein and his wife were continually fighting now about the children and howeager their finances seemed to be all the time. Albert was sure his marriage was over and began an affair with his cousin, Elsa.

At first, his wife accepted the list he gave her, but within three months she said enough was enough and Mileva and the boys returned to Zurich. Einstein was reported to have been standing on the platform waving goodbye and crying, the reason he was crying, no one knows. Because in just a few weeks he seemed to be over it and was as happy as a lark living alone in what he called "his tranquility."

❦

In 1918, while Pauline was visiting Maja, her daughter and Paul Winteler, Maja's husband in Luzern, they realized how ill Pauline had become and took her to Rosenau, a sanatorium where she could live and receive care.

❦

When Albert asked Mileva for the divorce, she had a nervous breakdown, and it seemed she was slow to recover.

❦

He had finally divorced Mileva in 1919 and agreed with her that if he ever won the Nobel Prize, he would give the money he might receive from it to her as part of the divorce settlement.

❦

At the end of 1919, Albert came and removed *his terminally ill mother* from the sanatorium and moved her in with Elsa and himself in Haberlandstrasse where she passed away a few months later in 1920. But, at least she was with family when she died.

❧

Then their son Eduard became a worry for him and Mileva. Eduard was very gifted. He was reading Schiller and Goethe in first grade and was blessed with a photographic memory. He could learn whatever he set his mind to at breathtaking speed. With all his intelligence Eduard was troubled. (Eduard had to finally be admitted into a mental facility due to schizophrenia, and *he died there in 1965*.)

❧

As Einstein had promised Mileva when **he won the Nobel Prize for Physics in 1922,** he would give the money from it to her. But he kept the award and gave proceeds equaling almost ten times what an average professor made to Maric. Maric was smart and invested it in three Zurich apartments, which she kept up until nearly the time of her death. (I have given you these two case scenarios of the money in the divorce settlement. Both came from reliable sources, and both seem plausible. However, from Albert's irrational acts at times, one tends to lean toward Albert losing most of it in the stock market crash.)

❧

Mileva devoted a lot of her time in taking care of Eduard and in 1947, she started to deteriorate healthwise. In 1948, she suffered from a stroke that resulted in paralysis on one side of her body. *Mileva died August 4, 1948*. Eduard, as said before died in 1965. One wonders if he was ever visited by his brother Hans Albert Einstein or his father, Albert Einstein.

❧

Elsa being Einstein's cousin caused everyone to talk, but Einstein did not care. Elsa's dad was Rudolf Einstein, and he was the "rich uncle."

◌⁂◌

Elsa had previously been married to a Max Loewenthal, from Berlin and they had two daughters, Margot and Ilse. They also had a son who had died shortly after birth.

◌⁂◌

Albert and Elsa moved in together in September 1917. Elsa was the one who kept putting the pressure on Albert to finalize the divorce with Maric.

◌⁂◌

Albert's main attraction with Elsa was that *she was a great cook*. He was also grateful to her for taking care of him when he was so ill with stomach issues. There was never any passion between the two of them, or that is what most thought. They married on June 2, 1919. Elsa was 43 and Einstein was 40.

◌⁂◌

Albert's life makes you wonder if he was a narcissist because his personal life seemed to always be in chaos and he was so aloof and callous. Elsa's daughter **Ilse** sent Dr. George Nicolai a close friend of hers a letter and told him when he read the letter to tear it up, but apparently, he did not, because it still exists.

◌⁂◌

Remember, Albert had been in the throes of a divorce from Mileva so he could marry Elsa, his cousin. Ilse was the oldest of Elsa's daughters, and she was working as Einstein's secretary.

❧

The letter was from Ilse, and it was a plea for advice. She told how she and Albert had been joking one day and all of sudden it got serious and Einstein asked her to marry him instead of her mother. She said that Albert told her he loved her, Ilse.

❧

And, crazy as it was, her mom, was willing to move over and let Ilse marry Einstein if that would make her happy.

❧

Albert was not going to make a choice, he said he would marry either one, he did not care. Ilse said that she knew he loved her very much, and probably more than any man, she would ever find because he told me so.

❧

Ilse did not feel like that about Albert. She thought of him as a father figure and did not want any physical relationship with him whatsoever.

❧

There's no evidence that Ilse and Albert ever consummated

their relationship. The next year Albert and Elsa were married and stayed married until *Elsa died in 1936*.

❦

Ilse married Rudolf Kayser who was a literary critic and a writer who eventually wrote a biography about Einstein. *In 1933, Ilse died of tuberculosis*.

III

EINSTEIN HAD A "MIRACLE YEAR"

"Imagination is more important than knowledge. Knowledge is limited. Imagination encircles the world."

ALBERT EINSTEIN

❧

When Einstein graduated in 1900, he faced possibly one of the biggest crises he had ever suffered. All this time he had studied and advanced at his "own" rate. He had always taken the liberty at cutting classes when he wanted to, and this angered some of his professors.

❧

One such professor happened to be **Weber** who Einstein needed a reference/letter of recommendation and Webber flat out refused to give him one because he had skipped so many classes. Due to this fact, every position that Einstein applied for, he was turned down. Einstein took the attitude that it was not fair and how could he have had a part in his bad luck. Imagine that; it sounds like Einstein had issues in taking responsibility, does it not?

❧

It was years that Einstein harbored the thought that his father had died thinking that he, Albert, was a failure. This thought would cause Albert extreme sadness when he would think about it. How could a man let his father die thinking his son was a failure? If only Einstein could have known what was in front of him?

❧

Albert had learned while at school that light speed is the same, so it does not matter how fast anyone moves. It made Newton's laws of motion a bunch of bolognas since Newton's Law had no absolute velocity.

❧

All this said and thought about made Einstein start to formulate a new principle of relativity: *Einstein's Principle:"the speed of light is a constant in any inertial frame (constantly moving frame)."*

❧

It was during 1905 that people often refer to as Einstein's **"Miracle Year"** because he came out with four published papers and every one of them would alter the course of our world and not just modern physics. They addressed fundamental problems about space, time, motion, matter, and the nature of energy.

❦

It was a viewpoint that was concerned with the transformation and production of "***Light and the Quantum Theory***" that was used by Einstein so he could explain the photoelectric effect. He explained that if light occurred in what we now know as **"photon"** or tiny packets, it should be able to take out the electrons found in metal.

❦

Einstein that same year offered for the first time experimental ***"proof that atoms"*** did exist. He did so by analyzing tiny particles that showed motion (Brownian movement) suspended in a glass of still water. By this finding, he had captured the ability to calculate sizes of jostling atoms known as "*Avogadro's number.*" It came to be known as the "*Movement of Small Particles Suspended in Stationary Liquids Required by Molecular-Kinetic Theory of Heat.*"

❦

When it came to the "*Electrodynamics of Moving Bodies,*" Einstein was on fire when he laid out his mathematical theory of "***special relativity***." With Special Relativity it can see the light as a field of continuous waves and not particles.

❦

Albert's brain was whirling overtime as it came out with "*Does *the Inertia of a Body Depend Upon Its Energy Content?*" Which almost got submitted as an afterthought. It was almost not presented, and that would have been a tragedy with him being so "on fire." The theory he almost forgot to submit revealed that the relativity theory was what led to the famous equation "$E=mc^2$."

⚜

It gave us our first way to explain the source of energy from the Stars and the Sun. This very equation could predict the evolution of power with almost a million times increased efficiency than that which was obtained by the ordinary physio-chemical methods. Even at first, Albert did not get the full picture of the implications of his formula, even though he was able to suggest the heat that was produced by radium could mark conversion of tiny amounts of the bulk of radium salts into energy.

⚜

Albert decided that he would submit a paper for his doctorate in 1905. Why not? It was such a Miracle Year, let's keep going!

⚜

Sure, there were other scientists out there like Lorentz and Poincare who had bits and pieces of the special *relativity theory*. But, it was Albert who held the Golden Key to the entire theory, and it was he that realized it was a universal law of nature, not some figment of the imagination or motion in the ether as the other two scientists wanted to believe.

Albert wrote in one of his letters to Mileva as he referred to "our theory," has made some wonder that maybe it was she who was the co-founder of the *theory of relativity*." Since Mileva had failed her graduate exams twice, she had decided to quit physics altogether. So, whether she helped or not will always be a mystery.

It is essential not to forget that during the 19th century there were two islands of thought when it came to physics: There were Newton's laws on motion, and there were Maxwell's ideas about light. Einstein had to take it upon himself in realizing and proving they contradicted each other and one of them had to be wrong.

It seemed at first that everyone was ignoring what Albert had to say in his papers. But, it all started changing when a physicist by the name of Max Planck who had discovered the quantum theory took notice.

It was in 1907 when Albert went face to face with the problem of gravitation. He started working with crucial insight into the fact that: "***gravity and acceleration are equivalent***," they were two angles of the exact phenomenon.

Albert even had minor work that resonated with the world. In 1910, he answered an age-old question as to "Why is the sky blue?" He wrote a paper explaining this phenomenon that was called "critical opalescence." It solved the problem as you examined the summed effect of the light scattering by its molecules in our atmosphere.

❧

In 1915, The General Theory of Relativity, which had taken Albert eight years working on the issue of gravity. In Albert's general relativity theory, he shows that *energy and matter will mold into the shape of the space and the flow of time*. What we can feel as gravity and its "force of pull" is simple. It is a sensation of following the most direct path through tortuous, four-dimensional space-time.It does sound like a radical vision:but space is no more in the same box than the universe comes in; what we find instead is: time and space, energy and matter are, as Einstein can prove, all locked in a most intimate embrace once you understand.

❧

It was shortly after Max Planck had commented on Albert's accomplishments that Einstein was being asked to speak at meetings internationally, like the Solvay Conference, and from there he rapidly rose in the world of academia.

❧

Talk about a Year of Miracle! He was offered positions at institutions of prestige such as the Swiss Federal Institute of Technology, University of Prague, University of Zurich, and the University of Berlin where he sat as the Director for the Kaiser Wilhelm Institute for Physics for twenty years.

ॐ

Albert's fame was spreading. He lived on the road, giving speeches at international affairs and lost in his deep thoughts of relativity.

❧ IV ❧
EVEN A GENIUS MAKES MISTAKES

"Try not to become a man of success, but rather try to become a man of value."
ALBERT EINSTEIN

❧

From 1905 to 1915, Albert was consumed with this nagging thought that there was a crucial problem, a possible flaw in his theory: he realized it had never made mention of acceleration or gravitation. He had a friend by the name of Paul Ehrenfest that had noticed if the disk is spinning, the outside rim will travel faster than it does in the center, so by special relativity you could place meter sticks on its circumference, and they should shrink. That would explain the Euclidian plane geometry had to fail for the disk.

❧

It would obsess Einstein's thoughts for the next ten years as he tried to develop **a theory of gravity** regarding when it came to the curvature of space-time. To Albert, Newton's idea of gravitational force seemed to be a by-product of a something much more profound: the bending of time and space.

⚜

Then came November 1915, Albert could finally take a deep breath as he felt he had completed the "*general theory of relativity*," which would leave a mark on the world as his masterpiece.

⚜

During the summer of that year, Albert gave two-hour lectures, six times at the University of Gottingen to thoroughly explain the version that he felt was complete on *general relativity*.

⚜

Then along came a real smack in the head when David Hilbert, a mathematician, that had put together the lectures for Albert at the University wrote a paper in November on the subject of *general relativity* five days before Albert. David Hilbert acted like it was indeed his work.He deserved to be tarred and feathered and driven out of town.

⚜

Albert and Hilbert did later patch up their quarrel and

remained friends. It seemed from letters that it was Einstein who put his right foot forward first.

> *He penned a letter to Hilbert saying: I struggled against a*
> *resulting sense of bitterness, and I did so with complete*
> *success. I once more thought of you in unclouded*
> *friendship and would ask you to try to do likewise*
> *toward me.*

Even today, some physicists refer to this action that the equations are obtained as the Einstein-Hilbert action, but, they know it solely belonged to Einstein.

SOMETIMES, THE MISTAKES COME BACK TO HAUNT THE FAMILY

❦

There is a girl, a granddaughter that died by the name of **Evelyn Einstein**. She was an adopted child at birth in 1941 by Hans and Frieda Einstein, Albert's son, and daughter-in-law.

❦

Evelyn said as a child that her parents told her that her real birth parents were her grandfather Albert Einstein and a ballet dancer with whom he had an affair.

❦

Evelyn only got to see her "grandfather" infrequently after her family moved to California and grandpa Albert lived in Princeton.

❦

Evelyn had no proof of anything of the sort, but in all the interviews she was subjected to she told the same story. She had been raised by Albert's son to save the entire Einstein family from being embarrassed.

❧

In another interview that she agreed to, she said: "I am outraged. It is hard for me to grasp that I would be treated as I have, which has been terrible."

❧

At one time, Evelyn had been married for 13 years to Grover Krantz, an eccentric anthropology professor from America who had become famous while trying to prove that *"Bigfoot"* did exist.

❧

After her marriage was over, she seemed to hit bottom, and became a 'dumpster diver,' looking for her next meal.

❧

When she died, she left behind no survivors. Money does not buy everything if you are never given any part of it or remembered for who you are in the family.

❧

Evelyn had been suffering for several years with diabetes, heart problems, and cancer. Evelyn said that Albert was never any "great being of science," to Evelyn he was just a plain old grandpa.

❦

Evelyn Einstein was intelligent; *she spoke five different languages*, and in Medieval Literature, she had a Master's degree. All she had done with her life was worked as a police officer, cult deprogrammer, and an animal control officer.

❦

Evelyn Einstein was homeless and had to live in her car for months after she went through a bitter divorce and died when she was 70 in Albany, California.

❦

Albert Einstein, one of the world's genius is considered by some to have been a, well, "pervert." Maybe one should say he enjoyed sexual conquests and thought about it all the time.

❦

I guess what caused him to be looked upon by some as a pervert was because he was so brazen about it with his wives. At one time he would have six girlfriends and would tell his wife how they were always showering him with what he called "unwanted" affection, according to some of his letters.

❦

It is hard to believe that it was "unwanted" affection on his part since he bragged about it so much. He wore all of it like a badge of honor, never giving it a second thought that someone may only be interested in him for his money or his fame and that in truth, his wrinkled clothes and unkempt appearance for most women would be a completeturn off?

Albert spent little time at home, but he was continually writing letters to his family. He would tell them all about his day, his discoveries and the people he had met and the new things he had seen.

The letters that were released in the 80s prove that when he was married to his second wife, who was his first cousin, he was cheating on her with his then secretary, Betty Neumann, and many others we are sure.

In his letters, Albert would tell about six women that he would spend time with and those who would bring him presents while he was married to his second wife.

These letters did not come to light until 1986 as his step-daughter Margot requested that they not be released until she had been dead for 20 years. One could not blame her.

Einstein identified that some of the women included an Ethel, Estella, a Toni and his "Russian Spy Lover," Margarita. He refers to some of the others only by their initials, such as L and M.

☙❧

Albert said in the letters that it was true M. had even followed him to England and the fact that her stalking me is getting way out of control. He went on to tell Margot that of all his dames, he was only attached to Mrs. L, and she is decent and harmless.

☙❧

He sent Margot another letter, and Albert asked her to pass on a note to Margarita, to stop providing inquisitive eyes with little tidbits.

☙❧

Barbara Wolff of the Hebrew University revealed that the M was a Berlin debutante *Ethel Michanowski* and she and Albert were involved during the 1920's and '30s. Wolff said it was an affair, but the only other information she would reveal was that M. was 15 years younger than Albert and was very helpful to his stepdaughters.

☙❧

In one of his letters to Elsa, he even opened up and said he thought that he was about fed up with his *"theory of relativity"* because when you are involved in something like that so much, even that interest will fade.

The newly found letters also tell the real story about Einstein's money from **the 1921 Nobel Prize**. There was that niggling little term of his divorce that the entire sum would be deposited into a Swiss bank account, and Maric could draw on the interest for her, Eduard, and Hans Albert to live on.

But what did Albert do? He invested most it in the United States and then lost most of it in the Depression. Go, Einstein! And, I mean GO Einstein! A man who has an IQ of 160 and does not evaluate the monetary situation before his investment, does leave one wondering. Did he not have a financial advisor? Why did he not obligate his agreement with his first wife?(This is the second version found in the research.)

It made Maric mad, and she felt betrayed, and she had every right to feel that way as Albert did not stick to his part of the bargain making her continually have to ask him for money.

In the end, he did give her more money than he had won with the Nobel Prize. The prize itself, in 1921, was worth $28,000, and in today's money, it would be $280,000.

Today, Albert's womanizing is over, but his image and name still draw almost ten million dollars a yearand all that goes to research and scholarships at the Israeli University.

EINSTEIN FINDS HE HAS ENEMIES

"Unthinking respect for authority is the greatest enemy of truth."

ALBERT EINSTEIN

❦❧

Einstein was a deep thinker, and that was no doubt. He was convinced that the theory of *general relativity* was right because it was so beautiful, and it could accurately predict Mercury's orbit around the Sun.

❦❧

In 1916, Einstein wrote and published *"Relativity, the Special and the General Theory: A Popular Exposition,"* 1920.

❦❧

His theory could predict by measure a deflection of light all around the Sun. Because of that fact, he offered to fund an entire expedition so that they could measure the deflection of starlight in the throes of a Solar eclipse. But Albert's work was to be interrupted by something called World War, and he was angry. Disgusted about the entire war, he called it *"the measles of mankind."*

⚜

He would go on to write, "At such a time as this, one realizes what a sorry species of animal one belongs to."

Einstein felt that racism of any kind was a disease.

⚜

After the war, chaos reigned in November 1918, and radical students took control of the University of Berlin, grabbed the college rector, several professors, and took them hostage. The college was worried that if they called the police, there would be a tragic result. But since Einstein was so respected by both faculty and students, he was the logical candidate to be called in as a mediator for the crisis. Einstein and Max Born did broker a compromise and resolved the issues.

⚜

Since Albert had always been a proponent of civil rights, he had objected to the way African Americans were treated. Einstein himself had suffered anti-Semitic discrimination in Germany before World War II, had worked with several civil rights activists and organizations and demanded they denounce segregation and racism and demanded equality.

There was an African-American civil rights supporter and singer by the name of Marian Anderson who was not being allowed a room at hotels and was being stopped from eating in public restaurants, so Einstein invited her to his house.

A bloody race riot broke out in 1946, and 500 state troopers carrying automatic weapons attacked and virtually destroyed every business owned by a black citizen within four blocks in Tennessee. 25 black men were jailed for attempted murder. Albert joined with Langston Huges, Thurgood Marshall, and Eleanor Roosevelt to fight for justice for the men. 24 of those 25 men were acquitted.

When two black couples in Monroe, Georgia were found murdered, and there was no justice served, Albert was furious; he wrote a letter to President Truman asking for prosecution of the lynchers and passage of a federal law for anti-lynching. Albert became friends with the actor Paul Robeson, and when he became blacklisted due to his activities against racism, Albert again opened his house to a friend of 20 years.

Einstein did not forget what he had wanted to do before the war. So as soon as he could, after the war, two expeditions left to test Albert's theory and his prediction about deflected starlight when near the Sun.

One of his groups sailed to the Island of Principe, which lay off the coast of Africa, while the other group was sent to Sobral in northern Brazil to observe the upcoming eclipse of the Sun on May 29, 1919.

Later that year on November 6th, the results were finally revealed in London to a joint meeting of the Royal Astronomical Society and the Royal Society itself.

The newspaper "The Times" of London headline read, "Revolution in Science – New Theory of the Universe – Newton's Ideas Overthrown – Momentous Pronouncement – Space 'Warped.'" Immediately, Albert Einstein was a household word and a world-renowned physicist.

The entire world wanted a part of Einstein, and in 1921, he started the first of many world tours in France, Japan, England, and even the United States.

It did not matter where he went, the crowds were wild and numbered into the thousands. While on his way to Japan he heard he had been given the Nobel Prize for Physics, but it was not for his relativity theory. Instead, it was for his photoelectric field. So, when he accepted the

Nobel Prize, he talked about relativity. That would show them.

❧

Einstein went so far as to become involved in cosmology. With his equations, he felt he could predict that our universe is dynamic – contracting and expanding. It was a direct contraindication to what the previous view had been that the universe was static.

❧

In 1928, Einstein was so overworked that he developed an enlarged heart that took him almost an entire year to recover from and be able to get back to his work.

❧

Einstein was able to find some recreation in playing his grand piano. A lot of his leisure time was also spent playing his violin. He would play in trios or quartets with his friends who were musically inclined to the fun.

❧

In 1929, after Albert turned 50 years old, he decided to build a summer house in Caputh where he would live with his little family every year starting in spring until the late part of autumn. He would do this until 1932.

❧

In 1929, the astronomer Edwin Hubble identified that our

universe was expanding, and this confirmed Einstein's work. Einstein made a visit in 1930 to Mount Wilson Observatory in California.

৩১৫৩

Hubble and Einstein met, and Albert told Hubble that the *cosmological constant* was his **"biggest mistake."** There is satellite data; recent that is, that has shown that this cosmological constant is probably not zero as thought. It seems to dominate the energy-matter content of our entire universe. It seems that Albert Einstein's "blunder," is the determinant of the ultimate fate of our universe.

৩১৫৩

At this period of his life, Einstein started corresponding with other "thinkers" that were influential during that time. Sigmund Freud (he and Einstein both had boys that had mental difficulties), and an Indian Mystic.

৩১৫৩

Albert started clarifying his views on religion, saying that he had the belief there was an "old one" who was the ultimate lawgiver. He went on to write that he, himself did not believe there was a personal God that would intervene in human affairs but did believe in the 17thcentury Dutch Jewish philosopher God.

৩১৫৩

To this Dutch Jewish philosopher God Albert said, was the God of beauty and harmony. Albert believed his task was to make up a master theory that allowed him to be able to "read God's mind."

⚜

Albert wrote,
I am not an atheist, and I do not think I can call myself a
pantheist. We are in the position of a little child entering a
huge library filled with books in many different
languages. The child dimly suspects a mysterious order in
the arrangement of the books but doesn't know what it
is. That, it seems to me, is the attitude of even the most
intelligent toward God.

⚜

All of Einstein's fame and theories, of course, led to a backlash. The backlash movement (Nazi) targeted "*relativity*," and branded it "Jewish physics," and they sponsored book burnings and conferences so that they could denounce Albert and all of his theories.

⚜

The Nazis didn't stop there but enlisted other Nobel laureates and physicists such as Philipp Lenard and Johannes Stark to try and make Einstein look like a fool.

⚜

In 1931, a book "*Authors Against Einstein*" was published.

When Albert was asked to comment on the book and what it
had said about him and his theory of relativity by all the

different scientists, Einstein simply replied, "to defeat
relativity one did not need the word of 100 scientists, just
one fact."

❧

In December 1932, Einstein without looking back left
Germany forever. He was moving and would never go back. It
was evident to him that his life was in danger if he stayed or
even visited there.

❧

A magazine had been published by a Nazi organization with
Einstein's picture that had the caption across it that said "*Not
Yet Hanged*" on the front cover. Albert found out that there
had been a price placed on his head. Due to the size of the
threat, Einstein had to re-examine his views on pacifism.

❧

Einstein came to America and settled himself in Princeton,
New Jersey at the new Institute for Advanced Study. It had
become the mecca for other physicists all over the world.

❧

Coming to America had been a challenge as well. When our
State Department got word that Albert Einstein was coming
to America to live, they proceeded to grill Albert on what his
political views were at that time. Finally, Einstein had a
meltdown.

His usual kind face became stern and his familiar melodious

voice loud, he cried out: "What is this, an inquisition? Is
this an attempt at deception? I do not plan to answer such
silly questions. I did not ask to go to America. Your people
invited me; yes, begged me. If I am to enter your country
as a suspect, I don't want to go at all. If you don't want to
give me a visa, then say so. Then I will know where
I stand."

◈

As soon as the press got their hands on this information, the State Department issued visas for Elsa and Albert the next day. They started their journey on December 10, 1932, for the United States and arrived at their destination on January 12th, 1933. It was only a couple of weeks later when Hitler took over Germany. It caused the Einsteins' to stay in the United States permanently.

◈

Newspapers trumpeted that the *"Pope of Physics"* had finally left Germany and Princeton was now his new Vatican.

ALBERT HAD HIS PROBLEMS TOO

"Great spirits have always encountered violent opposition from mediocre minds."
ALBERT EINSTEIN

❧☙

During the 1930's, Einstein faced some challenging years of his life. He suffered a lot of personal sorrow. His son, Eduard, suffered a nervous breakdown and was diagnosed with schizophrenia. As sad as it was, Eduard would remain in an institution for the rest of his life.

❧☙

Einstein had a close friend, another physicist by the name of Paul Ehrenfest, who had worked with him on the *general relativity theory* had become increasingly depressed and killed his fifteen-year-old son with Down's syndrome and then turned the pistol on himself in 1933. The reason he killed himself was not to be known until years later when his letters that he never mailed were discovered.

❦

Paul Ehrenfest's depression was so tremendous and the burden of his son's mental challenges and the cost of the institution's fees that would be left for his daughters to have to work all the time to pay were all too much, and Paul just snapped right there in the visitor's room of the institution. He pulled out a gun, killed his son and then turned the gun on himself.

❦

Einstein's second wife, Elsa, traveled to Paris to care for her dying daughter Ilse who had cancer. Her other daughter, Margot, came back home to be with her mother because Elsa, herself, had become very ill. She had developed kidney and heart problems, and on December 20, 1936, she died in the Einsteins' new Princeton home they had built in 1933.

❦

In the late 1930's, physicists were all looking to consider if, infact, their equations might be able to make an atomic bomb. Even in 1920, Einstein had considered making one but then decided against the idea. However, he left the design on the table if there could be a method found to multiplythe power of the atom.

❦

It finally happened in 1938-39 when Lise Meitner, Fritz Strassman, Otto Hahn, and Otto Frisch were able to show the vast amounts of energy one could release by splitting the atom of uranium. The news of this possibility created a fire in the physics society.

❧

It was in July 1939 when other physicists convinced Einstein he should be the one to write a letter to the United States President Franklin D. Roosevelt encouraging him to go ahead with developing an atomic bomb.

❧

Roosevelt wrote Einstein back, of course, letting Einstein know that he had in fact appointed a Uranium Committee to look at this issue.

❧

Einstein kept his Swiss Citizenship but became a man of two countries when he became an American Citizen in 1940.

❧

It was during the war that colleagues of Einstein's were asked to meet in Los Alamos; New Mexico to be a part of the Manhattan Project and make the first atomic bomb. It seemed to be one of the greatest ironies of Albert's career for the fact that Einstein the pacifist, through this one action, would help start an era of nuclear weaponry to the use he had always opposed. They never asked Albert to be a part of the effort even though it was his equation that launched the initiative.

❧

There are thousands of FBI files; they are now declassified, which explains the reason that Einstein was not a part of the making of the bomb. Albert being a part of the socialist and peace organizations, being the first and foremost reason for not allowing him to be part of this Project.

❧

J. Edgar Hoover wanted him to be tossed out of the United States entirely and use the Alien Exclusion Act to do so, but the U.S. State Department overruled Hoover and Albert remained in the United States.

❧

Hoover had all kinds of agents spying on poor Einstein. Hoover feared that intellectual, left-wing, pacifist Albert might be some threat to the U.S. establishment or who knows, maybe a Soviet Spy! Hoover even had them going through Albert's mail, listening in on his phone calls, and going through his trash for over 20 years. What a waste of taxpayers' money.

❧

Instead of being asked to leave, during the war, Einstein helped the U.S. Navy in evaluating some of their designs for weapon systems of the future. Einstein even aided the war effort monetarily by allowing some of his manuscripts to be auctioned.

❦

One handwritten copy of his paper written in 1905 on *special relativity* sold for $6.5 million. It can now be seen in the Library of Congress. 6.5 million dollars was a lot of money during that time of the world's history.

❦

When the atomic bomb was dropped on Japan, Einstein happened to be on vacation. When he heard the news, he joined the effort of bringing the use of the bomb under control by forming the "Emergency Committee of Atomic Scientists."

❦

The director of the atomic bomb project, Robert Oppenheimer had his security clearance taken away as he was suspected to have leftist associations.

❦

Then, Einstein started backing Oppenheimer and did not like the thought of developing the hydrogen bomb. He called for controls internationally on the technology of nuclear bombs. Einstein became ever more drawn into "anti-war activities and more and more involved in rights for the African American."

❦

It was in 1952 that David Ben-Gurion, the Premiere of Israel offered Einstein to be the President of Israel. Albert said that he was deeply moved by the fact that they had offered him this position, and at the same time he was ashamed and saddened that he could not accept. He went on to say that his entire life he had been dealing with scientific type matters, and he felt that he lacked what it took in natural aptitude and what it took to deal with people and to handle official functions. These facts alone are enough that I cannot fulfill what is required of this high office, even if I were getting older and my strength wasn't fading. My relationship with the Jewish nation is my strongest human bond, especially since I have become aware of how precarious it's situation is among all the countries of the entire world.

⚜

Einstein had a lot to say about God and his Hebrew heritage, and at times he made you think he did not believe in a God of any form.

> *Here is one of his quotes that is quite thought-provoking, "As a child, I received instruction both in the Bible and in the Talmud. I am a Jew, but I am enthralled by the luminous figure of the Nazarene... No one can read the Gospels without feeling the actual presence of Jesus. His personality pulsates in every word. No myth is filled with such life."*

⚜

Einstein loved the water and sailing, but he was not very good at it – he had neighbors on Long Island that could have told you how many times they had to rescue him when his sailboat would capsize. He had named the boat "Tinef" (a Yiddish word meaning "worthless"). As worthless as it was he sure liked playing with the sailboat and even worse, Albert never had learned how to swim! It was a good thing his neighbors watched out for him.

❦

And Einstein had a bad habit that did not help his health issues. He loved to smoke. It was in 1950 that he accepted a lifetime membership in the Pipe Smokers Club of Montreal. Einstein felt that smoking a pipe helped one to calm down and have objective judgment in human affairs.

❦

The doctors' placed smoking bans on Einstein all the time, but they did little good. When Einstein would walk around the Institute for Advanced Study at Princeton, Albert would pick up cigarette butts off the street and take out the discarded tobacco and fill up his pipe. At first, he walked across the meadow to the institute, but he found the street to have more leftover tobacco.Albert always wanted to get up the courage to defy the doctors' bans, but he didn't want to offend his friends.

❧ VII ❧

DID EINSTEIN HAVE A
₃RDSON? YOU DECIDE

Not everything that can be counted counts and not
everything that counts can be counted.
ALBERT EINSTEIN

❦

There was a weird letter which happened to turn up at the
Newspaper of "Evening Prague," one of Prague's largest, most
read, daily papers.

❦

*There was in the paper a note that stated: "I would like to
thank you and the country on behalf of the Einstein
family," is how the handwritten note started, "for the
celebration of the anniversary of my father's 100th
birthday. I am very grateful."*

❁

It had been signed by a man who said he was the only living child of Einstein. The letter at first was thought to be a joke or was it just another attempt for someone to try and immigrate to the United States, so it was dismissed.

❁

1980 rolled around, and all of Einstein's children had passed away. No one connected to the famous Mr. Albert Einstein, renowned physicist, who had lived in the city of Prague since 1912 when Albert himself had left for a more prestigious position in Zurich.

❁

Well, at the least it seems to be that it is what everyone seemed to think before Ludek Zakel Einstein came to the surface with a handful of papers and documents, a story of babies being switched at birth and a sworn statement of the dear woman who had raised him that he was Albert Einstein's son. He is about 63 years old and what else, but a physicist and he is working on projects basedon research from Albert's work. He looks so much like Einstein and has a temperament so like the man who had the ideas of our modern universe. Ludek has a young son that looks just like Albert.

❁

Zakel stated in a lengthy interview in his apartment that he may never be able to prove he is Albert's son, but he knows that he is. He goes on to say that he can't profit from the fact that he is Einstein's son and that he could only lose money. It would not change the course of hislife.So why would I want to lie about it?

✦

Ludek Zakel has worked all his life in Prague as a scientist, working to become a physicist way before 1972 when he states, Albert's step-daughter sent someone to inform him that his biological mother, had been Elsa. He had been raised by the woman who he had called mother because her child died the day before Ludek was born.

✦

Ludek's spare time is spent sculpting busts of Einstein and painting with watercolors of the same man as he tries to work out in his head the "significance of gravitational waves in space.

✦

He says he is more than ready to submit to DNA testing to prove what he is saying.

✦

This story that he keeps telling is all based on written but not authorized statements of two nurses that of course have been dead for years. There is the solemn vow of Margot, a possible half-sister, whose mother was Elsa; and the signed statement of 93-year-old Eva Zakel who will not speak about the whole matter anymore.(Elsa and Margot are both now deceased.)

✦

It is said that Mrs. Zakel did give birth to a baby that died on April 14, 1932, and she switched the child with the baby boy of Elsa Einstein who was 50 years old at the time. Elsa had come to Prague to see a doctor because she thought she had some tumor, but not pregnant.

❧

It seems that Zakel has an answer to every question that anyone can ask. Elsa had told her friends that she didn't want to find medical help there in Berlin at that time in 1932 because of the Nazis rising to power and Einstein and she was about to flee to America.

❧

The truth was, Albert did not want any more children, Elsa told her friends. Mrs. Zakel tells in her account that her baby boy died at birth and she was so desperate to make her husband happy because he wanted children. The original name picked out for the Zakel baby had been Jindrich,and in the hospital's log it is scratched out and replaced with Ludek. There has been no documentation ever found in any of the hospital logs of Elsa Einstein.

❧

Things like this could not happen today with tagging systems in hospitals and tagging parents to match their babies. But it was very possible63 years ago.

❧

The documents in existence – a birth certificate and the

baptismal records that were re-issued at the time of the Communist era that says Ludek is the son of Albert Einstein – well that could have been a clever ruse to escape their country.

⚜

Zakel had applied with the American Embassy more than once for citizenship.Every time he was turned down.

⚜

It will probably forever remain a mystery to the world since finding out Zakel has so much to lose if he is an Einstein. He has inherited buildings and land and if it should be proved he was Albert's son he could lose them all.

❧ VIII ❧

THE END IS SOON
TO COME

"Everybody is a genius. But if you judge a fish by its ability to climb a tree, it will live its whole life believing that it is stupid."
ALBERT EINSTEIN

❧

Einstein worked to develop more critical ideas towards his theory of *general relativity*. For instance, he wanted to prove things like wormholes, the possibility that there could be time travel, higher dimensions than we had ever dreamed, the creation of the universe, the existence of black holes and he seemed to become more isolated from the rest of the physics community all the time.

❧

The other physicists were working more on quantum theory, not relativity.

Einstein would often say of their ideas, "God does not play dice with the Universe."

☙❦❧

In 1935, it would happen to be the most celebrated in Albert's career on the quantum theory that led to the Einstein-Podolsky-Rosen "thought experiment."

☙❦❧

Under quantum theory, other individual circumstances could have two electrons that were separated by enormous distances could still have linked properties, like an umbilical cord.

☙❦❧

With these circumstances, if one were to measure the properties of the first electron, one would know the state of the second electron faster than light speed. Albert concluded that this violated relativity.

☙❦❧

There have been experiments since that time that have been able to prove that in fact, the quantum theory was correct and not Einstein.

☙❦❧

It seems that Einstein became more detached from his colleagues because he became so obsessed with discovering a theory to unify forces of our universe. It caused him in later years to stop opposing quantum theory and try to use it alongside gravity and light.

⚜

Einstein became so set in his ways. He quit traveling and just took long walks around the grounds of Princeton with some of his closest friends and associates with whom he could discuss religion, politics, unified field theory, and physics.

⚜

He was a regular sight to be seen walking around campus. His hair was always in disarray as if he never combed it or had just gotten out of bed. His pants were still wrinkled, and he usually wore a sweater. He never wore socks; they were a bother to him in the area where the big toe is located as they always had a big hole in that spot. Albert's second toe was much longer than his big toe next to it.

⚜

It was in 1950, he wrote and published an article in the Scientific American on his theory on strong force, but it was neglected by most, and it remained incomplete.

⚜

Albert seemed to believe that you must stand for what you believe in or you will fall for anything. He would never waver if his conscience told him to take action on a matter even if it was unpopular.

❦

One of these occasions was January 12, 1953, when Albert penned a letter to then-President Harry Truman.

> *It said: "My conscience compels me to urge you to commute the death sentence of Julius and Ethel Rosenberg."*

The two convicted atomic spies were executed five months later.

❦

Einstein's health had not always been so good. When he was 69, he went to his primary care physician and told him he had been having a lot of pain in the uppermost part of his stomach. He told him that off and on for several years he had been having attacks in his upper abdomen that would last at times for two to three days at a time and he almost always vomited when he had these attacks.

❦

He went on to add that this seemed to happen about every three to four months. He did smoke a pipe and seemed to be a bit overweight but not that much. When the doctor examined him, he could feel a mass deep down in the center of his stomach that was pulsating.

❦

Dr. Nissen, the very doctor who had developed the operation known as the *"Nissen Procedure,"* that prevents gastro-esophageal reflux, operated on Einstein by exploratory laparotomy at the Jewish Hospital in Brooklyn. When he opened Albert up, there lay an aortic aneurysm the size of a grapefruit.

ॐ

Nissen knew he could not ligate this large aneurysm and replacing the aorta with a graft was out of the question and yet to be perfected. All he could try to do was to reinforce the wall of the aorta and try to delay the inevitable rupture that would come.

ॐ

A good tissue irritant that produces marked fibrosis is poly-ethylene. So, Dr. Nissen while having Einstein open wrapped the anterior visible part of the aneurysm with this cellophane, hoping it would cause an intense fibrous reaction in the tissue, which would strengthen the wall of the aneurysm.

ॐ

Albert recovered in three weeks from the surgery during his hospital stay and returned to Princeton, New Jersey to his home.

ॐ

Off and on Albert would have some occasional back pain and would experience some pain much like gallbladder pain.

April 12, 1955, Einstein started having some pretty severe abdominal pain that got even worse the next day. Albert had a pretty good idea of what had happened and at first, refused to go to the hospital. He finally did go to the hospital because he felt he was a burden there at home. The Chief of Surgery at New York Hospital wanted to resect the aneurysm by a new procedure.

Albert refused the surgery and said that he wanted to go when he wanted. He felt you should not prolong your life artificially.Hefeltthat he had done his share on this earth and it was his time to go. By doing so, he would do it elegantly.

The night before Albert expired he had a view of his little round garden from the bed.

> *The nurse taking care of him asked, "Professor do you think God made the garden?"*
> *Einstein said, "Yes, God is both the gardener and the garden" to which the nurse replied, "Oh, I'd not thought of it that way" to which Einstein replied, "Yes, and I have spent my whole life just trying to catch a glimpse of Him at his work."*

He did, however, leave a piece of writing that happened to end unfinished. It happened to be his last words.

In essence, the conflict that exists today is no more than an
old-style struggle for power, once again presented to
mankind in semi-religious trappings. The difference is
that this time, the development of atomic power has
imbued the struggle with a ghostly character; for both
parties know and admit that should the quarrel
deteriorate into actual war, mankind is doomed. Despite
this knowledge, statesmen in responsible positions on both
sides continue to employ the well-known technique of
seeking to intimidate and demoralize the opponent by
marshaling superior military strength. They do so even
though such a policy entails the risk of war and doom.
Not one statesman in a position of responsibility has dared
to pursue the only course that holds out any promise of
peace, the course of supranational security, since for a
statesman to follow such a course would be tantamount to
political suicide. Political passions, once they have been
fanned into flame, exact their victims... Citater fra...

৩৩৩

Five days after being admitted to the hospital, Albert Einstein developed labored breathing and breathed his last *at 1:15 AM, April 18, 1955*. What a sad time for the history of the world.

৩৩৩

When he died, his body was moved from the hospital out to the funeral home before being cremated in Trenton. Most were not aware, but his brain did not get cremated, it went missing for many years. The pathologist of the hospital had confiscated it. It was finally found 23 years later by the journalist, Steven Levy who happened upon it pickled in a jar by one **Dr. Thomas Harvey** at Princeton where Einstein had expired.

❧

The rest of his body that was cremated; the ashes were spread around the Institute for Advanced Study in Princeton.

❧

Don't think that the pickling of his brain went to waste. There are recent studies that have revealed there were certain parts of Einstein's brain that were unusually convoluted. Not just that, his parietal lobes were "extraordinarily asymmetrical," and the motor cortices and somatosensory areas were "greatly" expanded in the left hemisphere.

❧

Further studies also show that Albert's brain cells had more of one type of brain cells called "glial" cells than our typical brains.

❧

Einstein's IQ was 160, but one has to wonder with the methods we have now of grading IQ, what his might genuinely rise to be.

❧ IX ❧

CHAPTER 8: WHAT EXACTLY WAS THE LEGACY OF EINSTEIN?

"Look deep into nature, and then you will understand everything better."
ALBERT EINSTEIN

❧

It seems as we look back that Einstein was far ahead of his time. His significant piece of unified field theory remained a total mystery in his lifetime.

❧

It was not until the 1970s and into the 80s that physicists started to unravel secrets associated with the *strong force* of the *quark model*.

❧

Even today Einstein's work allows other physicists to win Nobel Prizes as they build on his theories. In 1993 there was a Nobel Prize awarded to those who discovered gravitational waves that had been a prediction of Einstein's.

❦

Black holes now number in the thousands in space. The satellites we have in space have been able to verify precisely what Einstein spoke of in his lifetime.

❦

Even after Einstein retired,he kept working towards unifying the fundamental perceptions of physics, geometrization, to take the opposite approach, to the bulk of physicists.

❦

Everything that Einstein researched is well cataloged, and his most important works include The Special Theory of Relativity, General Theory of Relativity, Relativity Translations, The Evolution of Physics, About Zionism, My Philosophy, Why War?, Out of My Later Years, Investigations on Theory of Brownian Movement are some of the most important.

❦

Albert Einstein was the recipient ofhonorary doctorate degrees in medicine, philosophy, and science from universities in America and Europe. He was awarded the Franklin Medal from the Franklin Institute and the Copley Medal of the Royal Society of London.

❦

Einstein was always forward thinking, and he was not afraid to think of what would happen after his death.

❦

Notable for Einstein was that as he was nearing his last years of life, his views of God seemed to change.

One notices this in this quote: "As a child, I received instruction both in the Bible and in the Talmud. I am a Jew, but I am enthralled by the luminous figure of the Nazarene. No one can read the Gospels without feeling the actual presence of Jesus. His personality pulsates in every word. No myth is filled with such life."

One must admit this is in direct contrast as to how he felt when he was a hardened scientist.

❦

In his last will, he stated that everything he had, and all his intellectual property should be placed in a trust guarded by his stepdaughter and his secretary. His violin he left to his grandson. After his trustees had passed away, what remained of the rights of the trust were to be given to the Hebrew University in Jerusalem.

X

CONCLUSION

STRENGTHS

☙❧

A lbert Einstein had an IQ of 160 and who in the world would not love to have an IQ that high?

☙❧

Albert Einstein was liked well enough by professors and students alike that he could calm a terrible riot on the university where professors had been taken as hostages.

☙❧

Albert Einstein believed that if you could dream it and you worked toward it that you would eventually succeed in attaining the goal that you were reaching for. He set this example on many occasions.

WEAKNESSES

❧

Albert Einstein gave very little of himself to his family. It is the feelings of this author that he never really knew any of his children nor they him.

❧

Albert Einstein was never good to either of his wives. He was a braggart and boastful when it came to his conquests, and he did not care who in his family knew about the other women.

❧

Albert Einstein preferred being isolated from others rather than having to deal with other customs, opinions, and the prejudices of others.

❧

Albert Einstein, suffering from Asperger's Syndrome as a child, had so much to overcome as a child, and yet he forged on, and that alone was much to be admired. What he was able to do for the entire world was nothing short of amazing.

***

As I said in the book when he believed passionately in something and he did on many issues, he stood for what he believed in and stood firm. He did not let anyone talk him out of it.

***

He teaches us the reader to never give up on our dreams and that everyone is capable of reaching their goals no matter their humble beginnings.

***

Albert has many, many quotes that are highly thought-provoking and great to have nearby for reading.

INTERESTING BOOKS ON
ALBERT EINSTEIN YOU
MAY ENJOY:

Interesting Books on Albert Einstein you may enjoy:

৩১৯৩

Einstein's Tears

৩১৯৩

The Einstein File (From the FBI archives)

৩১৯৩

The World as I see It (by Einstein)

৩১৯৩

Albert Einstein – The Human Side

Copyright © 2019 by Kolme Korkeudet Oy

All rights reserved.

No part of this book may be reproduced in any form or by any electronic or mechanical means, including information storage and retrieval systems, without written permission from the author, except for the use of brief quotations in a book review.

YOUR FREE EBOOK!

As a way of saying thank you for reading our book, we're offering you a free copy of the below eBook.

Happy Reading!

GO WWW.THEHISTORYHOUR.COM/CLEO/

www.ingramcontent.com/pod-product-compliance
Lightning Source LLC
Chambersburg PA
CBHW030633220526
45463CB00004B/1510